T0301869

WATER HYDRAULICS

FUNDAMENTALS *for the* WATER & WASTEWATER MAINTENANCE OPERATOR SERIES

WATER HYDRAULICS

FRANK R. SPELLMAN, Ph.D.
JOANNE DRINAN

CRC Press
Taylor & Francis Group
Boca Raton London New York

CRC Press is an imprint of the
Taylor & Francis Group, an **informa** business

CRC Press
Taylor & Francis Group
6000 Broken Sound Parkway NW, Suite 300
Boca Raton, FL 33487-2742

© 2001 by Taylor & Francis Group, LLC
CRC Press is an imprint of Taylor & Francis Group, an Informa business

No claim to original U.S. Government works

ISBN-13: 978-1-56676-977-8 (hbk)

Visit the Taylor & Francis Web site at
http://www.taylorandfrancis.com

and the CRC Press Web site at
http://www.crcpress.com

Contents

4 PUMPING BASICS

5 FRICTION HEAD LOSS

6 BASIC PIPING HYDRAULICS

7 OPEN-CHANNEL FLOW

8 FLOW MEASUREMENT

9 HYDRAULIC MACHINES: PUMPS

10   FINAL REVIEW EXAMINATION 189

APPENDIX A

APPENDIX B

INDEX 209

Series Preface

Currently several books address broad areas of wastewater and waterworks operation. Persons seeking information for professional development in water and wastewater can locate study guides and also find materials on technical processes such as activated sludge, screening and coagulation. What have not been available until now are accessible treatments of each of the numerous specialty areas that operators must master to perform plant maintenance activities and at the same time to upgrade their knowledge and skills for higher levels of certification.

The *Fundamentals for the Water and Wastewater Maintenance Operator Series* is designed to meet the needs of operators who require essential background knowledge of subjects often overlooked or covered superficially in other sources. Written specifically for maintenance operators, the series comprises focused books designed to enhance knowledge and understanding.

Fundamentals for the Water and Wastewater Maintenance Operator Series covers over a dozen subjects in volumes that form stand-alone information guides or elements of a library of key topics. Areas to be presented in series volumes include: electricity, electronics, water hydraulics, water pumps, handtools, blueprint reading, piping systems, lubrication, and data collection.

Each volume in the series is written in a straightforward style without gon or complex calculations. All are heavily illustrated and include extensive, clearly outlined sample problems. Self-check tests are found within every chapter, and a comprehensive examination concludes each book.

The series provides operators with the information required for improved job performance. Equally important, using key points, worked problems, and sample test questions, the series is designed to help operators answer questions and solve problems on certification and licensure examinations.

Preface

Water/wastewater maintenance operators must be generalists, with a working knowledge on some level in several trades, sciences and disciplines. Does this mean that the water/wastewater maintenance operator must know everything about everything? Not at all. As a *generalist,* the fully competent maintenance operator must be knowledgeable in just about every aspect of water/wastewater maintenance and the ancillaries in science and engineering pertinent to the water/wastewater field—but only to a degree. This is especially true for workers in locations where maintenance operators are responsible for maintaining treatment unit process equipment as well as for operating and maintaining the plant.

If you accept the view that the water/wastewater maintenance operator must possess a wide-ranging knowledge of all aspects of water science along with standard maintenance skills—then a logical question might be: How is this possible? It is possible, though it isn't easy. Probably the most important step in the learning process is on-the-job training (OJT). However, many topics that make up the field of water/wastewater expertise not covered during OJT, or are unavailable through readily accessible forums. For years a gap has existed between the available training materials and the information water and wastewater operators need to know. *Fundamentals for the Water and Wastewater Maintenance Operator Series* is designed to bridge that gap. The present volume on water hydraulics bridges another gap in the available materials for operator training, the need for a clear presentation of hydraulics geared to plant operators.

But hydraulics is a broad field. We must first determine which branches of hydraulics apply to water and wastewater treatment. The maintenance operator is interested in two primary aspects of hydraulics: (1) hydraulics intended as a background necessary to understand the pumping and piping systems integral to water/wastewater treatment—water hydraulics; and (2) hydraulic principles used to power mechanical devices—fluid hydraulics. This volume covers *water hydraulics.* Fluid hydraulics (along with pneumatics) is addressed in a forthcoming volume of the series.

This volume, *Water Hydraulics,* covers those aspects of flow important to keeping water moving from one unit process to another, maintaining proper settling times, proper settling velocity, and providing lift to a higher elevation. The work focuses on the important principles main tenance operators need for working with water/wastewater hydraulics and associated appurtenances—concentrating on the practical side of the sub ject and stressing its usefulness as an important branch of water science and treatment.

This manual's goal is not to make hydraulic engineers out of any one—but instead to provide the basic principles required to understand the hydraulic process and equipment associated with it. To assure corre lation to modern practice and design, we present illustrative problems in terms of commonly used hydraulic parameters and cover typical hydraulic components (pumps, piping, etc.) found in today's water/wastewater treat ment systems.

Each chapter ends with a Self-Test to help you evaluate your mastery of the concepts we present. Before going on to the next chapter, take the Self-Test, compare your answers to the key, and review the pertinent in formation for any problems you missed. If you miss many items, review the whole chapter. A comprehensive final exam can be found at the end of this text.

 Note: *The symbol* ✔ *(check mark) displayed in various locations throughout this manual indicates a point is especially important* IMPORTANT *and should be studied carefully.*

This text is accessible to those who have no experience with water hydraulics; however, an understanding of basic math principles and how to use a calculator will help. If you work through the text systematically you will be surprised at how easily you acquire an understanding and skill in water hydraulics—adding a critical component to your professional knowledge.

Acknowledgements

To water and wastewater operators everywhere.

Water Hydraulics: What Is It?

Every man should keep minutes of what he reads . . . such an account would illustrate the history of his mind.

—Samuel Johnson

TOPIC

Setting the Stage

1.1 SETTING THE STAGE

The word "hydraulic" is derived from the Greek words "hydro" (meaning water) and "aulis" (meaning pipe). Originally, the term hydraulics referred only to the study of water at rest and in motion (flow of water in pipes or channels). Today, it is taken to mean the flow of *any* liquid in a system.

In terms of hydraulics, a liquid can be either oil or water. In fluid power systems used in modern industrial equipment, the hydraulic liquid of choice is oil. Some common examples of hydraulic fluid power systems include automobile braking and power steering systems, hydraulic elevators, and hydraulic jacks or lifts. The most familiar hydraulic fluid power systems in water/wastewater operations are used on dump trucks, front-end loaders, graders, and earth-moving and excavation equipment. Hydraulic fluid power systems, along with pneumatics (a fluid power system that uses gas as the transmitting fluid), are covered extensively in a forthcoming volume of the series on hydraulics and pneumatics. In this text, we are concerned with liquid water.

Many find the study of water hydraulics difficult and puzzling (especially the licensure examination questions), but we know it is not mysterious or difficult. It is the function or output of practical applications of the basic principles of water physics. This manual is another step in the direction of fulfilling a need for ground level, basic water/wastewater maintenance operations information, and/or to prepare you for more advanced water hydraulics training.

In *Water Hydraulics,* the traditional approach normally used to teach water hydraulics is abandoned. As with the other volumes in the series, the approach used is governed by the tenets of simplicity and factual clarity. This simple, clear approach presents only those areas actually needed for water hydraulics applications, rather than those found in traditional textbooks or training programs. The bottom line: in as simple a manner as possible, this manual covers the principles and calculations dealing with the hydraulics of water systems. It stresses only what is necessary for a basic understanding and emphasizes practical applications for water and wastewater maintenance operations.

The arrangement and approach of this manual has been tested on the deck plates—in the real world of water/wastewater treatment plant and collections operations. This material has been used effectively for several years to teach water hydraulics to maintenance operators who need formal training on the subject that is applicable to what they really do and are expected to know.

Because water/wastewater treatment is based on the principles of water hydraulics, concise, real-world training is required knowledge for those sitting for state licensure/certification examinations.

In today's modern world, water/wastewater maintenance operators who lack knowledge of basic water hydraulics are severely handicapped. Understanding water hydraulics is basic to water's movement (flow) through water and wastewater treatment systems from one unit process to the next.

We begin with hydraulics basics, including mathematical operations. Although math is an important part of water hydraulics, the math presented throughout this manual is fundamental level—only the operations that must be learned and nothing beyond the first-year algebra level. Each chapter is followed by a self-test that is used to gauge your progress through each lesson. A comprehensive final examination covering the material discussed in this manual is presented in the final chapter. This examination is an excellent way to measure your overall progress—it also serves to point out specific subject areas where refresher study may be beneficial.

This manual focuses on how to apply the few basic principles of modern water hydraulics practice. Properly presented, these principles are not difficult; instead they are accessible and essential for successful water/wastewater treatment plant operation.

Water Hydraulics: The Basics

Beginning students of water hydraulics and its principles often come to the subject matter with certain misgivings. For example, water/wastewater maintenance operators quickly learn on the job that their primary operational/maintenance concerns involve a daily routine of monitoring, sampling, laboratory testing, operation, and process maintenance. How does water hydraulics relate to daily operations? The hydraulic functions of the treatment process have already been designed into the plant. Why learn water hydraulics at all?

Simply put, while having hydraulic control of the plant site is obviously essential to the treatment process, maintaining and ensuring continued hydraulic control are also essential. No water/wastewater facility (and/or distribution-collection system) can operate without proper hydraulic control. The operator must know what hydraulic control is and what it entails to know how to ensure proper hydraulic control. Moreover, in order to understand the basics of piping and pumping systems, water/wastewater maintenance operators must have a fundamental knowledge of basic water hydraulics.

TOPICS

Basic Concepts
Properties of Water
Force and Pressure
Head
Flow/Discharge Rate: Water in Motion

Key Terms Used in This Chapter

HEAD — The measure of water pressure expressed as height of water in feet (1 psi = 2.31 feet of head). Stated another way, head is the equivalent distance water must be lifted to move from the supply tank or inlet to the discharge. Head can be divided into three components: static head, velocity head, and friction head.

FORCE — A push or pull influence that causes motion.

FRICTION HEAD — The energy needed to overcome friction in the piping system. It is expressed in terms of the added system head required.

STATIC HEAD — The actual distance from the system inlet to the highest discharge point.

VELOCITY HEAD — The energy needed to keep the liquid moving at a given velocity. It is expressed in terms of the added system head required.

TOTAL DYNAMIC HEAD — The total of the static head, friction head, and velocity head.

PRESSURE — The force exerted per square unit of surface area. Pressure = Weight × Height. In water, it is usually expressed as pounds per square inch (psi). One foot of water exerts a pressure of 0.433 pounds per square inch (psi).

VELOCITY — The speed of a liquid moving through a pipe, channel, or tank expressed in feet per second (fps).

HEADLOSS — The loss of energy commonly expressed in feet, as a result of friction. The loss is actually a transfer of heat.

2.1 INTRODUCTION[1]

Hydraulics—the study of fluids at rest and in motion—is essential for an understanding of how water/wastewater systems work, especially water distribution and wastewater collection systems. The principles of hydraulics discussed here will be useful in other chapters on flow, piping, and pumping systems. Specifically, this chapter (and subsequent chapters) covers those portions of water hydraulics that the operators of water/wastewater treatment facilities are likely to practice.

2.2 BASIC CONCEPTS

Air Pressure (@ Sea Level) = 14.7 pounds per square inch (psi)

The relationship shown above is important because our study of hydraulics begins with air. A blanket of air, many miles thick, surrounds the earth. The weight of this blanket on a given square inch of the earth's surface will vary according to the thickness of the atmospheric blanket above that point. As shown above, at sea level, the pressure exerted is 14.7 pounds per square inch (psi). On a mountaintop, air pressure decreases because the blanket is not as thick.

$1 \text{ ft}^3 \text{ H}_2\text{O} = 62.4 \text{ lb}$

The relationship shown above is also important: both cubic feet and pounds are used to describe a volume of water. There is a defined relationship between these two methods of measurement. The specific weight of water is defined relative to a cubic foot. One cubic foot of water weighs 62.4 pounds. This relationship is true only at a temperature of 4°C and at a pressure of one atmosphere [known as standard temperature and pressure (STP), 14.7 lb per square inch at sea level containing 7.48 gallons]. The weight varies so little that, for practical purposes, this weight is used

[1] Much of the information contained in this chapter is adapted from F. R. Spellman, *The Science of Water: Concepts and Applications.* Lancaster, PA: Technomic Publishing Co., Inc., 1998.

from a temperature 0°C to 100°C. One cubic inch of water weighs 0.0362 pounds. Water one foot deep will exert a pressure of 0.43 pounds per square inch on the bottom area (12 in. $\times$ 0.0362 lb/in.³). A column of water two feet high exerts 0.86 psi, one 10 feet high exerts 4.3 psi, and one 55 feet high exerts

$$55 \text{ ft} \times 0.43 \text{ psi/ft} = 23.65 \text{ psi}$$

A column of water 2.31 feet high will exert 1.0 psi. Producing a pressure of 50 psi requires a water column

$$50 \text{ psi} \times 2.31 \text{ ft/psi} = 115.5 \text{ ft}$$

IMPORTANT

Remember: *The important points being made here are*

1. *1 ft³ H₂O = 62.4 lb (see Figure 2.1)*

2. *A column of water 2.31 ft high will exert 1.0 psi*

   Another relationship is also important:

   1 gallon H₂O = 8.34 lb

At standard temperature and pressure, one cubic foot of water contains 7.48 gallons. With these two relationships, we can determine the weight of one gallon of water. This is accomplished by

wt. of gallon of water = 62.4 lb ÷ 7.48 gal = 8.34 lb/gal

Thus,

1 gallon H₂O = 8.34 pounds

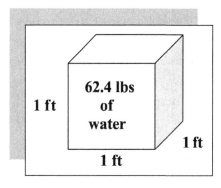

Figure 2.1
One cubic foot of water, resting on its bottom surface, exerts a force of 62.4 pounds on that square foot (62.4 lb/sq ft).

Further, this information allows cubic feet to be converted to gallons by simply multiplying the number of cubic feet by 7.48 gal/ft³.

Let's take a look at how we can put this information to work.

EXAMPLE 2.1

Problem: Find the number of gallons in a reservoir that has a volume of 855.5 ft³.

Solution:

$$855.5 \text{ ft}^3 \times 7.48 \text{ gal/ft}^3 = 6399 \text{ gallons (rounded)}$$

The term *head* is used to designate water pressure in terms of the height of a column of water in feet. For example, a 10-foot column of water exerts 4.3 psi. This can be called 4.3 psi pressure or 10 feet of head. (Note: we discuss head in greater detail in Section 2.5.)

Another example: If the static pressure in a pipe leading from an elevated water storage tank is 45 pounds per square inch (psi), what is the elevation of the water above the pressure gauge?

Remembering that 1 psi = 2.31 feet and that the pressure at the gauge is 45 psi,

$$45 \text{ psi} \times 2.31 \text{ ft/psi} = 104 \text{ ft (rounded)}$$

In demonstrating the relationship of the weight of water related to the weight of air, we can say theoretically that the atmospheric pressure at sea level (14.7 psi) will support a column of water 34 feet high:

$$14.7 \text{ psi} \times 2.31 \text{ ft/psi} = 34 \text{ ft (rounded)}$$

At an elevation of one mile above sea level, where the atmospheric pressure is 12 psi, the column of water would be only 28 feet high [12 psi × 2.31 ft/psi = 28 ft (rounded)].

If a glass or clear plastic tube is placed in a body of water at sea level, the water will rise in the tube to the same height as the water outside the tube. The atmospheric pressure of 14.7 psi will push down equally on the water surface inside and outside the tube.

However, if the top of the tube is tightly capped and all of the air is removed from the sealed tube above the water surface, forming a **perfect vacuum,** the pressure on the water surface inside the tube will be zero psi. The atmospheric pressure of 14.7 psi on the outside of the tube will push the water up into the tube until the weight of the water exerts the same 14.7 psi pressure at a point in the tube even with the water surface outside the tube. The water will rise 14.7 psi × 2.31 ft/psi = 34 feet.

In practice, it is impossible to create a perfect vacuum; thus, the water will rise somewhat less than 34 feet. The distance it rises depends on the amount of vacuum created. If, for example, enough air was removed from the tube to produce an air pressure of 9.7 psi above the water in the tube, how far will the water rise in the tube? To maintain the 14.7 psi at the outside water surface level, the water in the tube must produce a pressure of 14.7 psi − 9.7 psi = 5.0 psi. The height of the column of water that will produce 5.0 psi is

$$5.0 \text{ psi} \times 2.31 \text{ ft/psi} = 11.5 \text{ ft}$$

2.2.1 STEVIN'S LAW

Stevin's Law deals with water at rest. Specifically, the law states, "The pressure at any point in a fluid at rest depends on the distance measured vertically to the free surface and the density of the fluid." Stated as a formula, this becomes

$$p = w \times h \qquad\qquad (2.1)$$

where

p = pressure in pounds per square foot (psf)
w = density in pounds per cubic foot (lb/ft^3)
h = vertical distance in feet

EXAMPLE 2.2

Problem: What is the pressure at a point 18 feet below the surface of a reservoir?

Solution:

$$p = w \times h$$
$$= 62.4 \text{ lb/ft}^3 \times 18 \text{ ft}$$
$$= 1123 \text{ lb/ft}^2 \text{ or } 1123 \text{ psf}$$

IMPORTANT

Note: *To calculate this, we must know that the density of the water, w, is 62.4 pounds per cubic foot.*

Water/wastewater operators generally measure pressure in pounds per square **inch** rather than pounds per square **foot;** to convert, divide by 144 in²/ft² (12 in × 12 in = 144 in²):

$$P = \frac{1123 \ psf}{144 \text{ in}^2/\text{ft}^2} = 7.8 \text{ lb/in}^2 \text{ or psi (rounded)}$$

2.2.2 UNITS, CONVERSION FACTORS, AND FORMULAS

The measure of a counted quantity has a numerical value (for example, 6) and a unit (whatever there are 6 of). Examples of units are

● **distance**—inch, foot, yard, mile, millimeter, centimeter, meter, kilometer

● **mass**—ounce, pound, ton, milligram, gram, kilogram

● **time**—second, minute, hour, day

● **volume**—cubic feet, gallons, liters, cubic meters

● **concentration**—milligrams/liter, pounds per gallon

TABLE 2.1. Conversion Factors.

Multiply	By	To Obtain
Length		
inches	2.54	centimeters
inches	0.0254	meters
feet	0.3049	meters
yards	0.914	meters
miles	5280	feet
centimeters	10	millimeters
meters	100	centimeters
kilometers	0.62	miles
Area		
square foot	144	square inches
square yard	9	square feet
acres	43,560	square feet
square mile	640	acres
hectares	2.471	acres
hectares	107,600	square feet
Volume		
cubic feet	1728	cubic inches
cubic feet	7.48	gallons
cubic yards	27	cubic feet
millimeters	1	cubic centimeters
liters	1000	millimeters
gallons	3.785	liters
quart	946	milliliters
acre-inch	27,154	gallons
Weight		
pounds	453.6	grams
kilograms	2.206	pounds
ounces	28.3	grams
kilograms	1000	grams
grams	1000	milligrams
Time		
hours	3600	seconds
days	1440	minutes
Pressure		
atmospheres	29.92	inches of mercury
atmospheres	33.9	feet of water
atmospheres	760	millimeters of mercury
atmospheres	14.7	psi
feet of water	0.4335	psi
psi	2.307	feet of water

TABLE 2.1. (continued).		
Multiply	**By**	**To Obtain**
	Flow Rates	
cfs	0.6463	MGD
cfs	448.8	gpm
MGD	1.55	cfs
MGD	694	gpm
	Miscellaneous	
cubic feet of water	62.4	pounds
gallons of water	8.34	pounds
liters of water	1000	grams

We typically encounter both Metric System units and English System units in day-to-day plant operations.

Plant capacity in million gallons per day (MGD)
Residual chlorine levels in mg/L

English System units are older and sometimes awkward conversions: 12 inches = 1 foot, 3 feet = 1 yard and 5280 feet = 1 mile.

Metric System is newer, and conversions (based on the decimal system) are easier:1 meter = 39.37 inches and 1 pound = 454 grams.

$$55 \text{ inches} \times \frac{1 \text{ meter}}{39.37 \text{ inches}} = 1.4 \text{ meters (rounded)}$$

Note: See Table 2.1 for a more complete listing of conversion factors.

IMPORTANT

2.2.3 COMMONLY USED CONVERSION FACTORS (WATER/WASTEWATER OPERATIONS)

The following conversion factors are used extensively in waterworks operations and are commonly needed to solve problems on licensure examinations; the operator should keep them handy.

- 7.48 gallons per ft^3

- 3.785 liters per gallon

- 454 grams per pound

- 1000 mL per liter

- 1000 mg per gram

- 1 ft^3/sec (cfs) = 0.6465 MGD

IMPORTANT

Note: Density, also called specific weight, is mass per unit volume and may be registered as lb/cu ft, lb/gal, grams/mL, grams/cu meter. If we take a fixed volume container, fill it with a fluid, and weigh it, we can determine density of the fluid (after subtracting the weight of the container).

- **8.34 pounds per gallon (water)—(Density = 8.34 lb/gal)**

- **One milliliter of water weighs 1 gram—(Density = 1 gram/ml)**

- **62.4 pounds per ft^3 (water)—(Density = 8.34 lb/gal)**

- **8.34 lb/MG = mg/L** (converts dosage in mg/L into lb/day/MGD)
  Example: 1 mg/L × 10 MGD × 8.34 = 83.4 lb/day

- **1 psi = 2.31 feet of water (head)**

- **1 foot head = 0.433 psi**

- **°F = 9/5 (°C + 32)**

- **°C = 5/9 (°F − 32)**

- **Average Water Usage: 100 gallons/capita/day (gpcd)**

- **Persons per single family residence: 3.7**

2.2.4 COMMONLY USED BASIC FORMULAE (WATER/WASTEWATER OPERATIONS)

The following formulae are common operations used not only in waterworks operations but also in solving problems on licensure examinations.

💧 For a rectangle: Perimeter $= (2L) + (2W)$

                              Area $= LW$

💧 For a square: Perimeter $= 4S$

                              Area $= S^2$

💧 For a triangle: Perimeter $= S1 + S2 + S3$

                              Area $= (BH)/2$

💧 For a circle: Circumference $= 2r\pi$

                              Area $= \pi r^2 = \pi D^2/4$

                              diameter $= 2r$

                              π $= 3.14159 \ldots$ usually 3.14

                              π $=$ circumference/diameter

💧 For a cube: volume $= S^3$

💧 For a box: volume $= LWH$

💧 For a cylinder: volume $= \pi r^2 H = (\pi D^2 H)/4$

💧 Sphere: volume $= (4\pi r^3)/3 = (\pi D^3)/6$

💧 Triangular solid: volume $= (BHL)/2$

2.2.5 COMMONLY USED PRACTICAL FORMULAE (WATER/WASTEWATER OPERATIONS)

Area

Rectangular tank: $A = LW$

Circular tank: $A = \pi r^2$ or $A = 0.785 d^2$

Volume:

Rectangular tank: $V = LWH$

Circular tank: $V = \pi r^2 H$ or $0.785 d^2 H$

Flow:

Gal/day (gpd) = gal/min (gpm) $\times$ 1440 min/day

Gal/day (gpd) = gal/hr (gph) $\times$ 24 hr/day

Million gal/day (MGD) = (gal/day)/1,000,000

Dose:

lb = ppm $\times$ MG $\times$ 8.34 lb/gal

ppm = lb/(MG $\times$ 8.34 lb/gal)

Efficiency (% removal)

= influent − effluent/influent $\times$ 100

Weir loading (overflow rate)

$$= \frac{\text{total gallons/day}}{\text{length of weir}}$$

Surface settling rate

$$= \frac{\text{total gallons/day}}{\text{surface area of tank}}$$

Detention time (hours)

$$= \frac{\text{capacity of tank (gal)} \times 24 \text{ hr/day}}{\text{flow rate (gal/day)}}$$

Horsepower (hp)

$$= \frac{\text{gpm} \times \text{head (ft)}}{3960 \times \text{total efficiency}}$$

Pressure

= water is virtually incompressible; 62.4 lb of water occupies 1 cu ft

Specific Gravity

$$= \frac{\text{Density of Substance}}{\text{Density of Water}} = 1$$

therefore

$$= \frac{\text{Density of Water}}{\text{Density of Water}} = 1$$

The specific gravity of water is 1.

IMPORTANT

Key Point: *In real-world calculations and in the example problems to follow, it is important to ensure that unit dimensions are used in the same dimensions across the formula. It is easiest, for example, to label flow as cu ft/sec (cfs), area as sq ft, and velocity as ft/sec. If values are presented in other dimensions, first change them to one of these before inserting them into the formula.*

2.3 PROPERTIES OF WATER

Table 2.2 shows the relationship between temperature, specific weight, and density of water.

TABLE 2.2. Water Properties (Temperature, Specific Weight, and Density).		
Temperature (°F)	Specific Weight (lb/ft³)	Density (slugs/ft³)
32	62.4	1.94
40	62.4	1.94
50	62.4	1.94
60	62.4	1.94
70	62.3	1.94
80	62.2	1.93
90	62.1	1.93
100	62.0	1.93
110	61.9	1.92
120	61.7	1.92

TABLE 2.2. (continued).		
Temperature (°F)	Specific Weight (lb/ft³)	Density (slugs/ft³)
130	61.5	1.91
140	61.4	1.91
150	61.2	1.90
160	61.0	1.90
170	60.8	1.89
180	60.6	1.88
190	60.4	1.88
200	60.1	1.87
210	59.8	1.86

2.3.1 DENSITY AND SPECIFIC GRAVITY

When we say that iron is heavier than aluminum, this means that iron has greater density than aluminum. In practice, this means that a given volume of iron is heavier than the same volume of aluminum.

 Density is the mass per unit volume of a substance.

IMPORTANT

Suppose you had a tub of lard and a large box of cold cereal, each having a mass of 600 grams. The density of the cereal would be much less than the density of the lard because the cereal occupies a much larger volume than the lard occupies.

The density of an object can be calculated by using the formula:

$$\text{Density} = \frac{\text{Mass}}{\text{Volume}} \qquad (2.2)$$

In water/wastewater treatment, perhaps the most common measures of density are pounds per cubic foot (lb/ft³) and pounds per gallon (lb/gal).

🝆 **1 cubic foot of water weighs 62.4 lb—Density = 62.4 lb/cu /ft**

🝆 **One gallon of water weighs 8.34 lb—Density = 8.34 lb/gal**

The density of a dry material, such as cereal, lime, soda, and sand, is usually expressed in pounds per cubic foot. The density of a liquid, such as liquid alum, liquid chlorine, or water, can be expressed either as pounds per cubic foot or as pounds per gallon. The density of a gas, such as chlorine gas, methane, carbon dioxide, or air, is usually expressed in pounds per cubic foot.

As shown in Table 2.2, the density of a substance like water changes slightly as the temperature of the substance changes. This occurs because substances usually increase in volume (size—they expand) as they become warmer. Because of this expansion with warming, the same weight is spread over a larger volume, so the density is lower when a substance is warm than when it is cold.

Specific gravity is the weight (or density) of a substance compared to the weight (or density) of an equal volume of water. (Note: The specific gravity of water is 1).

IMPORTANT

This relationship is easily seen when a cubic foot of water, which weighs 62.4 lb as shown earlier, is compared to a cubic foot of aluminum, which weighs 178 pounds. Aluminum is 2.7 times as heavy as water.

It is not difficult to find the specific gravity of a piece of metal: weigh the metal in air, then weigh it under water. Its loss of weight is the weight of an equal volume of water. To find the specific gravity, divide the weight of the metal by its loss of weight in water.

$$\text{Specific Gravity} = \frac{\text{Weight of Substance}}{\text{Weight of Equal Volume of Water}} \qquad (2.3)$$

EXAMPLE 2.3

Problem: Suppose a piece of metal weighs 150 pounds in air and 85 pounds under water. What is the specific gravity?

Solution:

Step 1: 150 lb subtract 85 lb = 65 lb loss of weight in water

Step 2:

$$\text{specific gravity} = \frac{150}{65} = 2.3$$

Note: *In a calculation of specific gravity, it is essential that the densities be expressed in the same units.*

IMPORTANT

As stated earlier, the specific gravity of water is one (1), which is the standard, the reference to which all other liquid or solid substances are compared. Specifically, any object that has a specific gravity greater than one will sink in water (rocks, steel, iron, grit, floc, sludge). Substances with a specific gravity of less than one will float (wood, scum, gasoline). Considering the total weight and volume of a ship, its specific gravity is less than one; therefore, it can float.

The most common use of specific gravity in water/wastewater treatment operations is in gallons-to-pounds conversions. In many cases, the liquids being handled have a specific gravity of 1.00 or very nearly 1.00 (between 0.98 and 1.02), so 1.00 may be used in the calculations without introducing significant error. However, in calculations involving a liquid with a specific gravity of less than 0.98 or greater than 1.02, the conversions from gallons to pounds must consider specific gravity. The technique is illustrated in the following example.

EXAMPLE 2.4

Problem: There are 1455 gal of a certain liquid in a basin. If the specific gravity of the liquid is 0.94, how many pounds of liquid are in the basin?

Solution: Normally, for a conversion from gallons to pounds, we would use the factor 8.34 lb/gal (the density of water) if the substance's specific gravity was between 0.98 and 1.02. However, in this instance, the substance has a specific gravity outside this range, so the 8.34 factor must be adjusted.

Multiply 8.34 lb/gal by the specific gravity to obtain the adjusted factor:

Step 1: (8.34 lb/gal) (0.94) = 7.84 lb/gal (rounded)

Step 2: Then, convert 1455 gal to pounds using the corrected factor:

(1455 gal) (7.84 lb/gal) = 11,407 lb (rounded)

2.4 FORCE AND PRESSURE

Water exerts force and presssure against the walls of its container, whether it is stored in a tank or flowing in a pipeline. But there is a difference between force and pressure, although they are closely related. Force and pressure are defined below.

Force is the push or pull influence that causes motion. In the English system, force and weight are often used in the same way. The weight of a cubic foot of water is 62.4 pounds. The force exerted on the bottom of a one-foot cube is 62.4 pounds (see Figure 2.1). If we stack two cubes on top of one another, the force on the bottom will be 124.8 pounds.

Pressure is a force per unit of area. In equation form, this can be expressed as

$$P = \frac{F}{A} \qquad (2.4)$$

where

P = pressure

F = force

A = area over which the force is distributed

Earlier, we pointed out that pounds per square inch or pounds per square foot are common expressions of pressure. The pressure on the bottom of the cube is 62.4 pounds per square foot (see Figure 2.1). It is normal to express pressure in pounds per square inch (psi). This is easily accomplished by determining the weight of one square inch of a cube one foot high. If we have a cube that is 12 inches on each side, the number of square inches on the bottom surface of the cube is $12 \times 12 = 144$ in^2. Dividing the weight by the number of square inches determines the weight on each square inch.

$$\text{psi} = \frac{62.4 \text{ lb/ft}}{144 \text{ in.}^2} = 0.433 \text{ psi/ft}$$

This is the weight of a column of water one inch square and one foot tall. If the column of was two feet tall, the pressure would be 2 ft $\times$ 0.433 psi/ft = 0.866.

 Key Point: 1 foot of water = 0.433 psi

IMPORTANT

With the above information, feet of head can be converted to psi by multiplying the feet of head times 0.433 psi/ft.

EXAMPLE 2.5

Problem: A tank is mounted at a height of 90 feet. Find the pressure at the bottom of the tank.

Solution:

90 ft $\times$ 0.433 psi/ft = 39 psi (rounded)

To convert psi to feet, divide the psi by 0.433 psi/ft.

EXAMPLE 2.6

Problem: Find the height of water in a tank if the pressure at the bottom of the tank is 22 psi.

Solution:

$$\text{height in feet} = \frac{22\ psi}{0.433\ psi/ft}$$
$$= 51\ ft\ (rounded)$$

IMPORTANT

Important Point: *One of the problems encountered in a hydraulic system is storing the liquid. Unlike air, which is readily compressible and is capable of being stored in large quantities in relatively small containers, a liquid such as water cannot be compressed. Therefore, it is not possible to store a large amount of water in a small tank—62.4 lb of water occupies a volume of one cubic foot, regardless of the pressure applied to it.*

2.4.1 HYDROSTATIC PRESSURE

Figure 2.2 shows a number of differently shaped, connected, open containers of water. Note that the water level is the same in each container, regardless of the shape or size of the container. This occurs

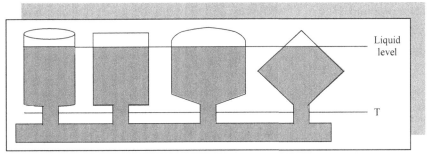

Figure 2.2
Hydrostatic pressure.

because pressure is developed, within water (or any other liquid), by the weight of the water above. If the water level in any one container were to be momentarily higher than that in any of the other containers, the higher pressure at the bottom of this container would cause some water to flow into the container having the lower liquid level. Also, the pressure of the water at any level (such as Line T) is the same in each of the containers. Pressure increases because of the weight of the water. The farther down from the surface, the more pressure is created. This illustrates that the **weight,** not the volume, of water contained in a vessel determines the pressure at the bottom of the vessel.

Some important principles always apply for hydrostatic pressure.[2]

1. The pressure depends only on the depth of water above the point in question (not on the water surface area).

2. The pressure increases in direct proportion to the depth.

3. The pressure in a continuous volume of water is the same at all points that are at the same depth.

4. The pressure at any point in the water acts in all directions at the same depth.

2.4.2 EFFECTS OF WATER UNDER PRESSURE[3]

Water under pressure and in motion can exert tremendous forces inside a pipeline. One of these forces, called hydraulic shock or *water hammer,* is the momentary increase in pressure that occurs when there is a sudden change of direction or velocity of the water.

[2] From Nathanson, J. A., *Basic Environmental Technology: Water Supply, Waste Management, and Pollution Control,* 2nd ed. Upper Saddle River, NJ: Prentice Hall, pp. 21–22, 1997.
[3] Adapted from information contained in Hauser, B. A., *Hydraulics for Operators.* Boca Raton, FL: Lewis Publishers, pp.16–18, 1993; *Basic Science Concepts and Applications: Principles and Practices of Water Supply Operations,* 2nd ed. Denver: American Water Works Association, pp. 351–353, 1995.

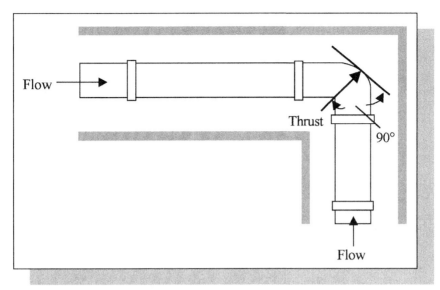

Figure 2.3
Shows direction of thrust in a pipe in a trench (viewed from above).

When a rapidly closing valve suddenly stops water flowing in a pipeline, pressure energy is transferred to the valve and pipe wall. Shock waves are set up within the system. Waves of pressure move in horizontal yo-yo fashion—back and forth—against any solid obstacles in the system. Neither the water nor the pipe will compress to absorb the shock, which may result in damage to pipes and valves and shaking of loose fittings.

Another effect of water under pressure is called thrust. *Thrust* is the force that water exerts on a pipeline as it rounds a bend. As shown in Figure 2.3, thrust usually acts perpendicular (at 90°) to the inside surface it pushes against. As stated, it affects bends, but also reducers, dead ends, and tees. Uncontrolled, the thrust can cause movement in the fitting or pipeline, which will lead to separation of the pipe coupling away from both sections of pipeline, or at some other nearby coupling upstream or downstream of the fitting.

There are two types of devices commonly used to control thrust in larger pipelines: thrust blocks and/or thrust anchors. A *thrust block* is a mass of concrete cast in place onto the pipe and around the outside bend of the turn. An example is shown in Figure 2.4. These are used for pipes with tees or elbows that turn left or right or slant upward. The

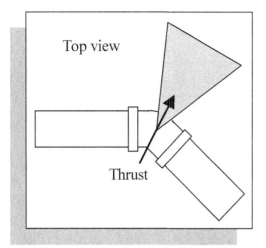

Figure 2.4
Thrust block.

thrust is transferred to the soil through the larger bearing surface of the block.

A *thrust anchor* is a massive block of concrete, often a cube, cast in place below the fitting to be anchored (see Figure 2.5). As shown in Figure 2.5, imbedded steel shackle rods anchor the fitting to the concrete block, effectively resisting upward thrusts.

The size and shape of a thrust control device depends on pipe size, type of fitting, water pressure, water hammer, and soil type.

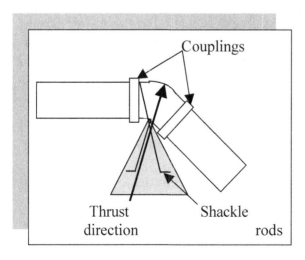

Figure 2.5
Thrust anchor.

2.5 HEAD

Head is defined as the vertical distance the water/wastewater must be lifted from the supply tank to the discharge or as the height a column of water would rise due to the pressure at its base. A perfect vacuum plus atmospheric pressure of 14.7 psi would lift the water 34 feet. If the top of the sealed tube is opened to the atmosphere and the reservoir is enclosed, then the pressure in the reservoir is increased, the water will rise in the tube. Because atmospheric pressure is essentially universal, we usually ignore the first 14.7 psi of actual pressure measurements and measure only the difference between the water pressure and the atmospheric pressure; we call this *gauge pressure*. For example, water in an open reservoir is subjected to the 14.7 psi of atmospheric pressure, but subtracting this 14.7 psi leaves a gauge pressure of 0 psi. This shows that the water would rise 0 feet above the reservoir surface. If the gauge pressure in a water main is 120 psi, the water would rise in a tube connected to the main:

$$120 \text{ psi} \times 2.31 \text{ ft/psi} = 277 \text{ ft} \qquad \text{(rounded)}$$

The *total head* includes the vertical distance the liquid must be lifted (static head), the loss to friction (friction head), and the energy required to maintain the desired velocity (velocity head).

Total Head = Static Head + Friction Head + Velocity Head

$$(2.5)$$

2.5.1 STATIC HEAD

Static head is the actual **vertical** distance the liquid must be lifted.

Static Head = Discharge Elevation − Supply Elevation

$$(2.6)$$

EXAMPLE 2.7

Problem: The supply tank is located at elevation 118 feet. The discharge point is at elevation 215 feet. What is the static head in feet?

Solution:

Static Head, ft = 215 ft − 118 ft = 97 ft

2.5.2 FRICTION HEAD

Friction head is the equivalent distance of the energy that must be supplied to overcome friction. Engineering references include tables showing the equivalent vertical distance for various sizes and types of pipes, fittings, and valves. The total friction head is the sum of the equivalent vertical distances for each component.

Friction Head, ft = Energy Losses Due to Friction *(2.7)*

2.5.3 VELOCITY HEAD

Velocity head is the equivalent distance of the energy consumed in achieving and maintaining the desired velocity in the system.

Velocity Head, ft = Energy Losses to Maintain Velocity

(2.8)

2.5.4 TOTAL DYNAMIC HEAD (TOTAL SYSTEM HEAD)

Total Head = Static Head + Friction Head + Velocity Head

(2.9)

2.5.5 PRESSURE/HEAD

The pressure exerted by water/wastewater is directly proportional to its depth or head in the pipe, tank, or channel. If the pressure is known, the equivalent head can be calculated.

$$\textbf{Head, ft} = \textbf{Pressure, psi} \times \textbf{2.31 ft/psi} \qquad (2.10)$$

EXAMPLE 2.8

Problem: The pressure gauge on the discharge line from the influent pump reads 72.3 psi. What is the equivalent head in feet?

Solution:

Head, ft = 72.3 × 2.31 ft/psi = 167 ft

2.5.6 HEAD/PRESSURE

If the head is known, the equivalent pressure can be calculated by

$$\textbf{Pressure, psi} = \frac{\textbf{Head, ft}}{\textbf{2.31 ft/psi}} \qquad (2.11)$$

EXAMPLE 2.9

Problem: The tank is 22 feet deep. What is the pressure in psi at the bottom of the tank when it is filled with water?

Solution:

$$\text{Pressure, psi} = \frac{22\text{ ft}}{2.31\text{ ft/psi}}$$
$$= 9.52\text{ psi (rounded)}$$

2.6 FLOW/DISCHARGE RATE: WATER IN MOTION

The study of fluid flow is much more complicated than that of fluids at rest, but it is important to have an understanding of these principles because the water in a waterworks and distribution system and in a waste-water treatment plant and collection system is nearly always in motion.

Discharge (or flow) is the quantity of water passing a given point in a pipe or channel during a given period of time. Stated another way for open channels: The flow rate through an open channel is directly related to the velocity of the liquid and the cross-sectional area of the liquid in the channel.

$$Q = A \times V \tag{2.12}$$

where

Q = Flow—discharge in cubic feet per second (cfs)
A = Cross-sectional area of the pipe or channel (ft^2)
V = water velocity in feet per second (fps or ft/sec)

EXAMPLE 2.10

Problem: The channel is 6 feet wide, and the water depth is 3 feet. The velocity in the channel is 4 feet per second. What is the discharge or flow rate in cubic feet per second?

Solution:

$$\text{Flow, cfs} = 6 \text{ ft} \times 3 \text{ ft} \times 4 \text{ ft/second}$$
$$= 72 \text{ cfs}$$

Discharge or flow can be recorded as gallons/day (gpd), gallons/minute (gpm), or cubic feet (cfs). Flows treated by many waterworks or wastewater treatment plants are large and often are referred to in million gallons per day (MGD). The discharge or flow rate can be converted from cfs to other units such as gallons per minute (gpm) or million gallons per day (MGD) by using appropriate conversion factors (see Table 2.1).

EXAMPLE 2.11

Problem: A pipe 12 inches in diameter has water flowing through it at 10 feet per second. What is the discharge in (a) cfs, (b) gpm, and (c) MGD? Before the basic formula (2.12) can be used, the area of the pipe must be determined. The formula for the area of a circle is (π is the constant value 3.14159 or simply 3.14)

$$A = \pi \times \frac{D^2}{4} = \pi \times r^2 \qquad (2.13)$$

where

D = diameter of the circle in feet
r = radius of the circle in feet

Therefore, the area of the pipe is

$$A = \pi \frac{D^2}{4} = 3.14 \times \frac{(1 \text{ ft})^2}{4} = 0.785 \text{ ft}^2$$

Now, the discharge in cfs [part (a)]can be determined:

$$Q = V \times A = 10 \text{ ft/sec} \times 0.785 \text{ ft}^2$$
$$= \textbf{7.85 } \text{ft}^3\text{/sec or cfs}$$

For part (b), 1 cubic foot per second is 449 gallons per minute (see Table 2.1), so 7.85 cfs × 449 gpm/cfs = **3520 gpm.**

Finally, for part (c), one million gallons per day is 1.55 cfs (see Table 2.1), so

$$\frac{7.85 \text{ cfs}}{1.55 \dfrac{\text{cfs}}{\text{MGD}}} = \textbf{5.06 MGD}$$

IMPORTANT

Important Point: *Flow may be laminar (streamline) (see Figure 2.6) or turbulent (see Figure 2.7). Laminar flow occurs at extremely low velocities. The water moves in straight parallel lines, called streamlines, or laminae, which slide upon each other as they travel, rather than mixing up. Normal pipe flow is turbulent flow, which occurs because of friction encountered on the inside of the pipe. The outside layers of flow are thrown into the inner layers; the result is that all the layers mix and are moving in different directions and at different velocities. However, the direction of flow is forward.*

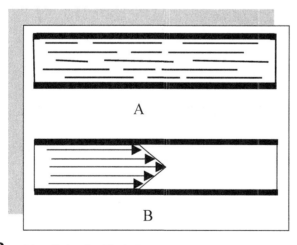

A

B

Figure 2.6
Laminar (streamline) flow.

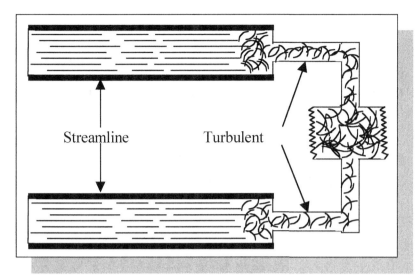

Figure 2.7
Turbulent flow.

Important Point: *Flow may be steady or unsteady. For the purposes of this book, only steady state flow is considered; that is, most* IMPORTANT *of the hydraulic calculations in this manual assume steady-state flow.*

2.6.1 AREA/VELOCITY

The *law of continuity* states that the discharge at each point in a pipe or channel is the same as the discharge at any other point (if water does not leave or enter the pipe or channel). That is, under the assumption of steady-state flow, the flow that enters the pipe or channel is the same flow that exits the pipe or channel. In equation form, this becomes

$$Q_1 = Q_2 \text{ or } A_1V_1 = A_2V_2 \qquad (2.14)$$

Important Note: *In regards to the area/velocity relationship, Equation (2.13) also makes clear that, for a given flow rate,* IMPORTANT *the velocity of the liquid varies indirectly with changes in cross-sectional area of the channel or pipe. This principle provides the basis for many of the flow measurement devices used in open channels (weirs, flumes, and nozzles).*

EXAMPLE 2.12

Problem: A pipe 12 inches in diameter is connected to a 6-inch diameter pipe. The velocity of the water in the 12-inch pipe is 3 fps. What is the velocity in the 6-in. pipe?

Solution: Using the equation $A_1V_1 = A_2V_2$, we need to determine the area of each pipe:

$$\text{12 in.: } A = \pi \times \frac{D^2}{4}$$

$$= 3.14 \times \frac{(1\ ft)^2}{4}$$

$$= 0.785\ ft^2$$

$$\text{6 in.: } A = 3.14 \times \frac{(0.5)^2}{4}$$

$$= 0.196\ ft^2$$

The continuity equation now becomes

$$(0.785\ ft^2) \times \left(3\frac{ft}{sec}\right) = (0.196\ ft^2) \times V_2$$

Solving for V_2

$$V_2 = \frac{(0.785\ ft^2) \times (3\ ft/sec)}{(0.196\ ft^2)}$$

$$= 12\ ft/sec \text{ or fps}$$

2.6.2 PRESSURE/VELOCITY

In a closed pipe flowing full (under pressure), the pressure is indirectly related to the velocity of the liquid. This principle, when combined with the principle discussed in the previous section, forms the basis for several flow measurement devices (Venturi meters and rotameters) as well as the injector used for dissolving chlorine into water, and chlorine, sulfur dioxide, and/or other chemicals into wastewater.

$$\textbf{Velocity}_1 \times \textbf{Pressure}_1 = \textbf{Velocity}_2 \times \textbf{Pressure}_2 \qquad (2.15)$$

or

$$V_1P_1 = V_2P_2$$

REFERENCES

Basic Science Concepts and Applications: Principles and Practices of Water Supply Operations Series, 2nd ed. Denver: American Water Works Association, 1995.

Hauser, B. A., *Hydraulics for Operators.* Boca Raton, FL: Lewis Publishers, 1993.

Nathanson, J. A., *Basic Environmental Technology: Water Supply, Waste Management, and Pollution Control,* 2nd ed. Upper Saddle River, NJ: Prentice Hall, 1997.

Spellman, F. R., *The Science of Water: Concepts and Applications.* Lancaster, PA: Technomic Publishing Co. Inc., 1998.

Self-Test

IMPORTANT

Note: *Answers to chapter self-tests are found in Appendix A.*

2.1 Find the number of gallons in a storage tank that has a volume of 660 ft^3.

2.2 Suppose a rock weighs 160 pounds in air and 125 pounds under water. What is the specific gravity?

2.3 There are 1450 gal of a certain liquid in a storage tank. If the specific gravity of the liquid is 0.91, how many pounds of liquid are in the tank?

2.4 A tank is mounted at a height of 85 feet. Find the pressure at the bottom of the tank.

2.5 Find the height of water in a tank if the pressure at the bottom of the tank is 16 psi.

2.6 The elevation of the liquid in the supply tank is 2666 ft. The elevation of the liquid surface of the discharge is 2130 ft. What is the total static head of the system?

2.7 The fluid in a fluid power system can be either a(n) _____ or a(n) _____.

2.8 A push or pull applied against an object to move it is called a(n) _____.

2.9 The density of a liquid is expressed in terms of _____.

2.10 The specific gravity of a liquid is determined by comparing the weight of the fluid to the weight of an equal volume of _____ at the same temperature.

2.11 The weight of liquid contained in a vessel determines the _____ at the bottom of the vessel.

2.12 The water in a tank weighs 910 pounds. How many gallons does it hold?

2.13 A liquid with a specific gravity of 1.10 is pumped at a rate of 40 gpm. How many pounds per day are being delivered by the pump?

2.14 The pressure gauge at bottom of a standpipe reads 115 psi. What is the depth of water in the standpipe?

2.15 A 110-ft diameter cylindrical tank contains 1.6 MG water. What is the water depth?

2.16 The pressure in a pipeline is 6400 psf. What is the head on the pipe?

2.17 The pressure on a surface is 35 psig. If the surface area is 1.6 sq ft, what is the force (lb) exerted on the surface?

Bernoulli's
Theorem

. . . They will take your hand and lead you to the pearls of the desert, those secret wells swallowed by oyster crags of wadi, underground caverns that bubble rusty salt water you would sell your own mothers to drink.[4]

TOPICS

Conservation of Energy
Piezometric Surface
Bernoulli's Equation

3.1 INTRODUCTION

To keep plant systems operating properly and efficiently, an understanding of the basics of hydraulics is needed—the laws of force, motion, and others. As stated previously, most applications of hydraulics in water/wastewater treatment systems involve water in motion—in pipes under pressure or in open channels under the force of gravity. The volume of water flowing past any given point in the pipe or channel per unit time is called the flow rate or discharge—or just *flow*.

In regards to flow, *continuity of flow* and the *continuity equation* have been discussed [i.e., Equation (2.14)]. Along with the continuity of flow principle and continuity equation, the law of conservation of energy, piezometric surface, and Bernoulli's Theorem (or Principle) are also important to our study of water hydraulics. These important principles are discussed in this chapter.

[4]Holman, S. *A Stolen Tongue*. New York: Anchor Press, Doubleday, p. 245, 1998.

Key Terms Used in This Chapter

CONSERVATION OF ENERGY	A basic principle in physics that energy can neither be created nor destroyed, but it can be converted from one form to another.
PIEZOMETRIC SURFACE	An imaginary surface that coincides with the level of the water in an aquifer or the level to which water in a system would rise in a piezometer.
PIEZOMETER	An instrument for measuring pressure head in a conduit or tank, by determining the location of the free water surface.
ELEVATION HEAD	Pressure due to the elevation of the water.
PRESSURE HEAD	The height of a column of water that a given hydrostatic pressure in a system could support (see Section 2.5.5).
VELOCITY HEAD	A measurement of the amount of energy in water due to its velocity, or motion.
HYDRAULIC GRADE LINE (HGL)	A line (hydraulic profile) indicating the piezometric level of water at all points along a conduit, open channel, or stream. In an open channel, the HGL is the free water surface.

3.2 CONSERVATION OF ENERGY

Many of the principles of physics are important to the study of hydraulics. When applied to problems involving the flow of water, few of the principles of physical science are more important and useful than the *Law of Conservation of Energy*. Simply, the Law of Conservation of Energy states that energy can neither be created nor destroyed, but it can be converted from one form to another. In a given closed system, the total energy is constant.

3.2.1 ENERGY HEAD

Two types of energy, kinetic and potential, and three forms of mechanical energy exist in hydraulic systems: potential energy due to elevation, potential energy due to pressure, and kinetic energy due to velocity. Energy has the units of foot pounds (ft-lb). It is convenient to express hydraulic energy in terms of *Energy Head,* in feet of water. This is equivalent to foot-pounds per pound of water (ft lb/lb = ft).

3.3 PIEZOMETRIC SURFACE[5]

The preceding chapter discussed that when a vertical tube, open at the top, is installed onto a vessel of water, the water will rise in the tube to the water level in the tank. The water level to which the water rises in a tube is the *piezometric surface.* That is, the piezometric surface is an imaginary surface that coincides with the level of the water to which water in a system would rise in *a piezometer* (an instrument used to measure pressure).

The surface of water that is in contact with the atmosphere is known as *free water surface.* Many important hydraulic measurements are based on the difference in height between the free water surface and some point in the water system. The **piezometric surface** is used to locate this free water surface in a vessel, where it cannot be observed directly.

To understand how a piezometer actually measures pressure, consider the following example.

If a clear, see-through pipe is connected to the side of a clear glass or plastic vessel, the water will rise in the pipe to indicate the level of the water in the vessel. Such a see-through pipe, the piezometer, allows you to see the level of the top of the water in the pipe; this is the piezometric surface.

In practice, a piezometer is connected to the side of a tank or pipeline. If the water-containing vessel is not under pressure (as is the case in

[5] From Spellman, F. R., *The Science of Water: Concepts & Applications.* Lancaster, PA: Technomic Publishing Co., Inc., pp. 92–93, 1998.

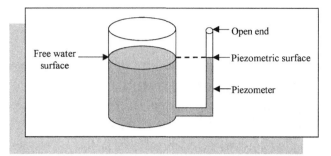

Figure 3.1
*A container not under pressure where the piezometric surface
is the same as the free water surface in the vessel.*

Figure 3.1), the piezometric surface will be the same as the free water
surface in the vessel, just as it would if a drinking straw (the piezometer)
were left standing in a glass of water.

When pressurized in a tank and pipeline system, the pressure will
cause the piezometric surface to rise above the level of the water in the
tank. The greater the pressure, the higher the piezometric surface (see
Figure 3.2). An increased pressure in a water pipeline system is usually
obtained by elevating the water tank.

IMPORTANT

Note: *In practice, piezometers are not installed on pipelines or
water towers because water towers are hundreds of feet high.
Instead, pressure gauges are used that record pressure in feet of
water or in psi.*

Water only rises to the water level of the main body of water when
it is at rest (static or standing water). The situation is quite different when

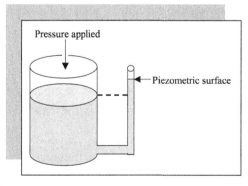

Figure 3.2
*A container under pressure
where the piezometric surface is
above the level of the water in
the tank.*

water is flowing. Consider, for example, an elevated storage tank feeding a distribution system pipeline. When the system is at rest and all valves are closed, all the piezometric surfaces are the same height as the free water surface in storage. On the other hand, when the valves are opened and the water begins to flow, the piezometric surface changes. This is an important point because as water continues to flow down a pipeline, less and less pressure is exerted. This happens because some pressure is lost (used up), keeping the water moving over the interior surface of the pipe (friction). The pressure that is lost is called *head loss.*

3.3.1 HEAD LOSS

Head loss is best explained by example. Figure 3.3 shows an elevated storage tank feeding a distribution system pipeline. When the valve is closed [Figure 3.3(a)], all the piezometric surfaces are the same height as the free water surface in storage. When the valve opens and water begins to flow [Figure 3.3(b)], the piezometric surfaces **drop.** The farther along the pipeline, the lower the piezometric surface, because some of the pressure is used up keeping the water moving over the rough interior surface of the pipe. Thus, pressure is lost and is no longer available to push water up in a piezometer; this is the head loss.

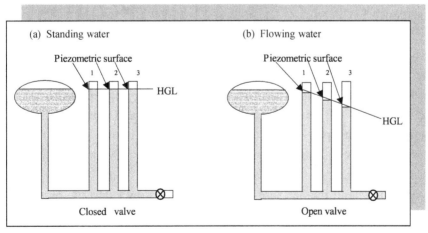

Figure 3.3
Shows head loss and/or piezometric surface changes when water is flowing.

3.3.2 HYDRAULIC GRADE LINE (HGL)

When the valve is opened in Figure 3.3, flow begins with a corresponding energy loss due to friction. The pressures along the pipeline can measure this loss. In Figure 3.3(b), the difference in pressure heads between sections 1, 2, and 3 can be seen in the piezometer tubes attached to the pipe. A line connecting the water surface in the tank with the water levels at section 1, 2, and 3 shows the pattern of continuous pressure loss along the pipeline. This is called the *Hydraulic Grade Line (HGL)* or *Hydraulic Gradient* of the system. (Note: It is important to point out that in a static water system, the HGL is always horizontal.) The HGL is a very useful graphical aid when analyzing pipe flow problems.

Key Point: *Changes in the piezometric surface occur when water is flowing.*

IMPORTANT

3.4 BERNOULLI'S THEOREM[6]

Swiss physicist and mathematician Samuel Bernoulli developed the calculation for the total energy relationship from point to point in a steady-state fluid system in the 1700s. Before discussing Bernoulli's energy equation, it is important to understand the basic principle behind Bernoulli's equation.

Water (and any other hydraulic fluid) in a hydraulic system possesses two types of energy—kinetic and potential. *Kinetic energy* is present when the water is in motion. The faster the water moves, the more kinetic energy is used. *Potential energy* is a result of the water pressure. The *total energy* of the water is the sum of the kinetic and potential energy. Bernoulli's principle states that the total energy of the water (fluid) always remains constant. Therefore, when the water flow in a system increases, the pressure must decrease. When water starts to flow in a hydraulic system, the pressure drops. When the flow stops, the pressure

[6] Adapted from Nathanson, J. A., *Basic Environmental Technology: Water Supply, Waste Management, and Pollution Control,* 2nd ed. Upper Saddle River, NJ: Prentice Hall, pp. 29–30, 1997

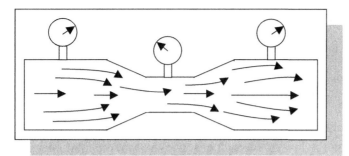

Figure 3.4
Demonstrates Bernoulli's principle.

rises again. The pressure gauges shown in Figure 3.4 indicate this balance more clearly.

 Important Point: *The basic principle explained above ignores friction losses from point to point in a fluid system employing* IMPORTANT *steady-state flow.*

3.4.1 BERNOULLI'S EQUATION

In a hydraulic system, total energy head is equal to the sum of three individual energy heads. This can be expressed as

Total Head = Elevation Head + Pressure Head + Velocity Head

where

elevation head—pressure due to the elevation of the water
pressure head—the height of a column of water that a given hydrostatic pressure in a system could support
velocity head—energy present due to the velocity of the water

This can be expressed mathematically as

$$E = z + \frac{P}{w} + \frac{v^2}{2g} \qquad (3.1)$$

where

E = total energy head
z = height of the water above a reference plane, ft

p = pressure, psi (must be converted to lb/ft^2, see Note in Example 3.1)

w = unit weight of water, 62.4 lb/ft^3

v = flow velocity, ft/s

g = acceleration due to gravity, 32.2 ft/s^2

Consider the constriction in the section of pipe shown in Figure 3.5. We know, based on the law of energy conservation, that the total energy head at section A, E_1, must equal the total energy head at section B, E_2, and using Equation (3.1), we get Bernoulli's equation.

$$z_A + \frac{P_A \times 144}{w} + \frac{v_A^2}{2g} = z_B + \frac{P_B \times 144}{w} + \frac{V_B^2}{2g} \qquad (3.2)$$

The pipeline system shown in Figure 3.5 is horizontal, therefore, we can simplify Bernoulli's equation because $z_A = z_B$.

Because they are equal, the elevation heads cancel out from both sides, leaving:

$$\frac{P_A}{w} + \frac{v_A^2}{2g} = \frac{P_B}{w} + \frac{v_B^2}{2g} \qquad (3.3)$$

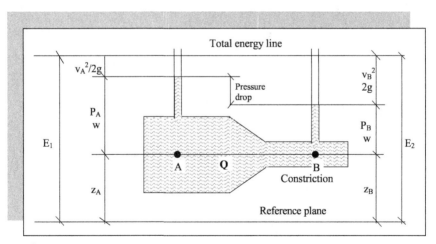

Figure 3.5

Shows the result of the law of conservation. Because the velocity and kinetic energy of the water flowing in the constricted section must increase, the potential energy may decrease. This is observed as a pressure drop in the constriction. (Adapted from Nathanson, 1997, p. 29.)

As water passes through the constricted section of the pipe (section B), we know from continuity of flow that the velocity at section B must be greater than the velocity at section A, because of the smaller flow area at section B. This means that the velocity head in the system increases as the water flows into the constricted section. However, the total energy must remain constant. For this to occur, the pressure head and, therefore, the pressure must drop. In effect, pressure energy is converted into kinetic energy in the constriction.

The fact that the pressure in the narrower pipe section (constriction) is less than the pressure in the bigger section seems to defy common sense, but it does follow logically from continuity of flow and conservation of energy. The fact that there is a pressure difference allows measurement of flow rate in the closed pipe. This phenomenon is included in the discussion of flow measurement.

EXAMPLE 3.1

Problem: In Figure 3.5, the diameter is 8 in. at section A and 4 in. at section B. The flow rate through the pipe is 3.0 cfs, and the pressure at section A is 100 psi. What is the pressure in the constriction at section B?

Solution:

Step 1: Compute the flow area at each section, as follows:

$$A_A = \frac{\pi(0.666 \text{ ft})^2}{4} = 0.349 \text{ ft}^2 \ (rounded)$$

$$A_B = \frac{\pi(0.333 \text{ ft})^2}{4} = 0.087 \text{ ft}^2$$

Step 2: From $Q = A \times V$ or $V = Q/A$, we get

$$V_A = \frac{3.0 \text{ ft}^3/s}{0.349 \text{ ft}^2} = 8.6 \text{ ft/s} \ (rounded)$$

$$V_B = \frac{3.0 \text{ ft}^3/\text{s}}{0.087 \text{ ft}^2} = 34.5 \text{ ft/s (rounded)}$$

Step 3: Applying Equation (3.3), we get

$$\frac{100 \times 144}{62.4} + \frac{8.6^2}{2 \times 32.2} = \frac{p_B \times 144}{62.4} + \frac{34.5^2}{2 \times 32.2}$$

Continuing, we get

$$231 + 1.15 = 2.3p_B \times 18.5$$

and

$$p_B = \frac{232.2 - 18.5}{2.3} = \frac{213.7}{2.3}$$
$$= 93 \text{ psi (rounded)}$$

Note: *The pressures are multiplied by 144 in.^{2}/ft^2 to convert from psi to lb/ft^2 to be consistent with the units for w; the energy head terms are in feet of head.*

IMPORTANT

REFERENCES

Holman, S., *A Stolen Tongue*. New York: Anchor Press, Doubleday, p. 245, 1998.

Nathanson, J. A., *Basic Environmental Technology: Water Supply, Waste Management, and Pollution Control*. Upper Saddle River, NJ: Prentice Hall, pp. 29–30, 1997.

Spellman, F. R., *The Science of Water*. Lancaster, PA: Technomic Publishing Co., Inc., pp. 92–93, 1998.

Self-Test

3.1 Bernoulli's principle states that the total energy of a hydraulic fluid is _____ _____.

3.2 What is *pressure head?*

3.3 Briefly describe what is meant by *continuity of flow.*

3.4 What is a *hydraulic grade line?*

3.5 A flow of 1500 gpm takes place in a 12-inch pipe. Calculate the velocity head.

3.6 Water flows at 5.00 mL/sec in a 4-inch line under a pressure of 110 psi. What is the pressure head (ft of water)?

3.7 In question 3.6, what is the velocity head in the line?

3.8 What is the velocity head in a 6-inch pipe connected to a 1-ft pipe, if the flow in the larger pipe is 1.46 cfs?

Pumping Basics

Only the sail can contend with the pump for the title of the earliest invention for the conversion of natural energy to useful work, and it is doubtful that the sail takes precedence. Since the sail cannot, in any event, be classified as a machine, the pump stands essentially unchallenged as the earliest form of machine that substituted natural energy for muscular effort in the fulfillment of man's needs.[7]

TOPICS

Pumps: Classification and Characteristics
Pumping Hydraulics
Hydraulics of Wells and Wet Wells

Key Terms Used in This Chapter

DYNAMIC PUMPS	Pumps in which the energy is added to the water continuously and the water is not contained in a set volume.
POSITIVE DISPLACEMENT PUMPS	Pumps in which the energy is added to the water periodically and the water is contained in a set volume.
DRAWDOWN (IN WELL)	The distance between the static level and the pumping level.
CONE OF DEPRESSION	The depression, rough conical in shape, produced in a water table or other piezometric surface by the extraction of water from a well at a given rate.

[7]Krutzsch, W. C., Introduction and Classification of Pumps. In *Pump Handbook,* Karassik, I. J. et al. New York: McGraw-Hill Inc., p. 1–1, 1976.

4.1 INTRODUCTION[8]

Conveying water/wastewater to and from process equipment is an integral part of the water/wastewater industry that requires energy consumption. The amount of energy required depends on the height to which the water/wastewater is raised, the length and diameter of the conveying conduits, the rate of flow, and the water/wastewater's physical properties (in particular, viscosity and density). In some applications, external energy for transferring water/wastewater is not required. For example, when water/wastewater flows to a lower elevation under the influence of gravity, a partial transformation of the water/wastewater's potential energy into kinetic energy occurs. However, when conveying water or wastewater through horizontal conduits, especially to higher elevations within a system, mechanical devices such as pumps are employed. Requirements vary from small units used to pump only a few gallons per minute to large units capable of handling several hundred cubic feet per second.

 Note: In determining the amount of pressure or force a pump must provide to move the water or wastewater, the term **pump** IMPORTANT **head** was established.

Several methods are available for transporting water, wastewater, and chemicals for treatment between process equipment:

- centrifugal force inducing fluid motion

- volumetric displacement of fluids, either mechanically or with other fluids

- transfer of momentum from another fluid

- mechanical impulse

- gravity induced

Depending on the facility and unit processes contained within, all of the methods above may be important to the maintenance operator.

[8] Adapted from Cheremisinoff, N. P. and Cheremisinoff, P. N., *Pumps/Compressors/Fans: Pocket Handbook.* Lancaster, PA: Technomic Publishing Co., Inc., p. 3, 1989.

However, emphasis in this chapter is placed on the basics of the mechanical devices for conveying water/wastewater, namely, pumps.

4.2 PUMPS: CLASSIFICATION AND CHARACTERISTICS

Before discussing the classification and characteristics of the pumps used in water/wastewater operations, it is important to understand exactly what a pump is. For practical purposes, a pump is a hydraulic machine that transforms the mechanical energy generated by a motor into the energy of moving fluids.

4.2.1 PUMP TYPES

The pumps used in water/wastewater systems can be divided into two general categories: dynamic pumps and displacement pumps (see Figure 4.1).

Note: The basic difference between the dynamic and displacement pump has to do with their response to changes in discharge pressure.

IMPORTANT

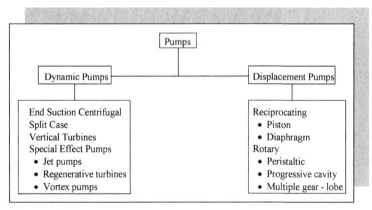

Figure 4.1
Types of pumps.

One type of dynamic pump, the centrifugal pump, is the most common pump used in water/wastewater systems. Displacement pumps are also used and are commonly called positive displacement pumps. The most common positive displacement pump is the diaphragm pump used to pump (meter) chemical into a unit process (e.g., chlorine, sodium hypochlorite, fluoride solutions, and others).

4.2.1.1 DYNAMIC PUMPS

Dynamic pumps are used in conditions where high volumes are required and a change in flow is not a problem (see Table 4.1). As the discharge pressure on a dynamic pump is increased, the quantity of water pumped is reduced.

TABLE 4.1. Pump Applications in Water/Wastewater Systems.*		
Application	Function	Pump Type
Low service	To lift water from the source to treatment processes, or from storage to filter-backwashing system	Centrifugal
High service	To discharge water under pressure to distribution system; to pump collected or intercepted wastewater to treatment facility	Centrifugal
Booster	To increase pressure in the distribution/collection system or to supply elevated storage tanks	Centrifugal
Well	To lift water from shallow or deep wells and discharge it to the treatment plant, storage facility, or distribution system	Centrifugal or jet
Chemical feed	To add chemical solutions at desired dosages for treatment processes	Positive displacement
Sampling	To pump water/wastewater from sampling points to the laboratory or automatic analyzers	Positive displacement or centrifugal
Sludge/biosolids	To pump sludge or biosolids from sedimentation facilities to further treatment or disposal	Positive displacement or centrifugal
*Adapted from *Water Transmission and Distribution,* 1996, p. 358.		

According to Garay (1990), dynamic pumps have "been developed in historically recent times. It may be said, however, that a person swinging a bucket filled with water, throwing the water into an elevated receiver, is using a dynamic pumping principle. Commonly, however, when motion is imparted to a fluid—usually a rotary motion—the increase in potential energy due to the fluid motion may be converted into a higher degree of pressure, concurrent with the transfer of the fluid. This, then, is the dynamic principle of pumping."[9]

Important Point: *Dynamic pumps can be operated for short periods of time with the discharge valve closed.*

IMPORTANT

4.2.1.2 POSITIVE DISPLACEMENT PUMPS

Positive displacement pumps are used primarily in conditions where relatively small volumes of a precise amount are required. In operation, the fluid is forced to move because the movement of a piston, vane, roller, or screw displaces it. Positive displacement pumps act to force water into a system regardless of the resistance that may oppose the transfer. They will not change their volume with a change in discharge pressure.

Caution: *Positive displacement pumps must never be operated when the discharge valve is closed; otherwise, the pump may be*
IMPORTANT *damaged.*

4.3 PUMPING HYDRAULICS[10]

During operation, water enters a pump on the suction side, where the pressure is lower. Since the function of the pump is to add pressure to the system, discharge pressure will always be higher. An important concept

[9] Garay, P. N., *Pump Application Desk Book.* Lilburn, GA: The Fairmont Press (Prentice Hall), p. 10, 1990.
[10] Adapted from Arasmith, S., *Introduction to Small Water Systems.* Albany, OR: ACR Publications, Inc., pp. 59–61, 1993.

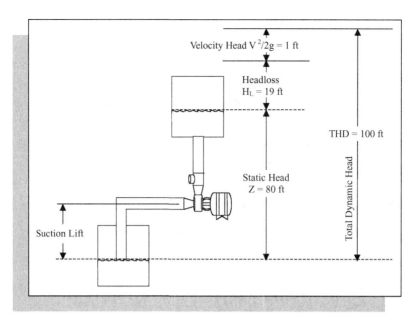

Velocity Head $V^2/2g = 1$ ft

Headloss $H_L = 19$ ft

THD = 100 ft

Static Head $Z = 80$ ft

Total Dynamic Head

Suction Lift

Figure 4.2
Components of Total Dynamic Head.

to keep in mind: in pump systems, measurements are taken from the point of reference to the centerline of the pump (horizontal line drawn through center of pump).

To understand pump operation, or *pumping hydraulics,* we need to be familiar with certain basic terms and then relate these terms pictorially (as we do in Figure 4.2) to illustrate how water is pumped from one point to another.

● *Static Head*—the distance between the suction and discharge water levels when the pump is shut off. We indicate static head conditions with the letter *Z* (see Figure 4.2).

● *Suction Lift*—the distance between the suction water level and the center of the pump impeller. This term is only used when the pump is in a suction lift condition; the pump must have the energy to provide this lift. A pump is said to be in a suction lift condition any time the center (eye) of the impeller is above the water being pumped (see Figure 4.2).

● *Suction Head*—a pump is said to be in a suction head condition any time the center (eye) of the impeller is below the water level being pumped. Specifically, suction head is the distance between the suction water level and the center of the pump impeller when the pump is in a suction head condition (see Figure 4.2).

● *Velocity Head*—the amount of energy required to bring water or wastewater from standstill to its velocity. For a given quantity of flow, the velocity head will vary indirectly with the pipe diameter. Velocity head is often shown mathematically as $V^2/2g$ (see Figure 4.2).

● *Total Dynamic Head*—the total energy needed to move water from the center line of a pump (eye of the first impeller of a lineshaft turbine) to some given elevation or to develop some given pressure. This includes the static head, velocity head and the headloss due to friction (see Figure 4.2).

4.4 HYDRAULICS OF WELLS AND WET WELLS

When the source of water for a water distribution system is from a groundwater supply, knowledge of well hydraulics is important to the operator. Basic well hydraulics terms are presented and defined, and they are related pictorially (see Figure 4.3). Also discussed are wet wells, which are important, both in water and wastewater operations.

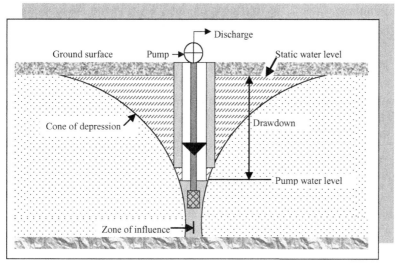

Figure 4.3
Hydraulic characteristics of a well.

4.4.1 WELL HYDRAULICS

● *Static Water Level*—the water level in a well when no water is being taken from the groundwater source i.e., the water level when the pump is off; see Figure 4.3). Static water level is normally measured as the distance from the ground surface to the water surface. This is an important parameter because it is used to measure changes in the water table.

● *Pumping Water Level*—the water level when the pump is off. When water is pumped out of a well, the water level usually drops below the level in the surrounding aquifer and eventually stabilizes at a lower level, which is the pumping level (see Figure 4.3).

● *Drawdown*—the difference, or the drop, between the static water level and the pumping water level, measured in feet. Simply, it is the distance the water level drops once pumping begins (see Figure 4.3).

● *Cone of Depression*—in unconfined aquifers, there is a flow of water in the aquifer from all directions toward the well during pumping. The free water surface in the aquifer then takes the shape of an inverted cone or curved funnel line. The curve of the line extends from the pumping water level to the static water level at the outside edge of the zone (or radius) of influence (see Figure 4.3).

 Note: *The shape and size of the cone of depression is dependent on the relationship between the pumping rate and the rate at* **IMPORTANT** *which water can move toward the well. If the rate is high, the cone will be shallow, and its growth will stabilize. If the rate is low, the cone will be sharp and continue to grow in size.*

● *Zone (or Radius) of Influence*—the distance between the pump shaft and the outermost area affected by drawdown (see Figure 4.3). The distance depends on the porosity of the soil and other factors. This parameter becomes important in well fields with many pumps. If wells are set too close together, the zones of influence will overlap, increasing the drawdown in all wells. Obviously, pumps should be spaced far enough apart to prevent this from happening.

Two important parameters not shown in Figure 4.3 are well yield and specific capacity. *Well yield* is the rate of water withdrawal that a well can supply over a long period of time, or, simply, the maximum pumping rate that can be achieved without increasing the drawdown. The yield of small wells is usually measured in gallons per minute (liters per minute) or gallons per hour (liters per hour). For large wells, it may be measured in cubic feet per second (cubic meters per second).

Specific capacity is the pumping rate per foot of drawdown (gpm/ft), or

$$\text{Specific Capacity} = \text{Well Yield} \div \text{Drawdown} \qquad (4.1)$$

EXAMPLE 4.1

Problem: If the well yield is 300 gpm and the drawdown is measured to be 20 ft, what is the specific capacity?

Solution:

Specific Capacity = 300 ÷ 20

Specific Capacity = 15 gpm per ft of drawdown

Specific capacity is one of the most important concepts in well operation and testing. The calculation should be made frequently in the monitoring of well operation. A sudden drop in specific capacity indicates problems such as pump malfunction, screen plugging, or other problems that can be serious. Such problems should be identified and corrected as soon as possible.

4.4.2 WET WELL HYDRAULICS

Water pumped from a wet well by a pump set above the water surface exhibits the same phenomena as the groundwater well. In operation, a slight depression of the water surface forms right at the intake line (drawdown), but in this case it is minimal because there is free water at the pump entrance at all times (at least there should be). The most important consideration in wet well operations is to ensure that the suction line is submerged far enough below the surface, so that air entrained by the active movement of the water at this section is not able to enter the pump.

Because water or wastewater flow is not always constant or at the same level, variable speed pumps are commonly used in wet well operations, or several pumps are installed for single or combined operation. In many cases, pumping is accomplished in an on/off mode. Control of pump operation is in response to water level in the well. Level control devices such as mercury switches are used to sense a high and low level in the well and transmit the signal to pumps for action.

REFERENCES

Arasmith, S., *Introduction to Small Water Systems.* Albany, OR: ACR Publications, Inc., pp. 59–61, 1993.

Cheremisinoff, N. P. and Cheremisinoff, P. N., *Pumps/Compressors/Fans: Pocket Handbook.* Lancaster, PA: Technomic Publishing Co., Inc., p. 3, 1989.

Garay, P. N., *Pump Application Desk Book.* Lilburn, GA: The Fairmont Press (Prentice Hall), p. 10, 1990.

Krutzsch, W. C., Introduction and Classification of Pumps. In *Pump Handbook,* Karassik, I. J. et al. New York: McGraw-Hill Inc., p. 1–1, 1976.

Water Transmission and Distribution, 2nd ed. Denver: American Water Works Association, p. 358, 1996.

Self-Test

4.1 What are the two major categories of pumps?

4.2 The energy placed into the water by a pump can be expressed as an increase in flow and an increase in _____.

4.3 What term is used to describe the difference between the level of water in a well and the level of water in the reservoir when the pump is shut down?

4.4 When water is drawn out of a well, a _____ of _____ will develop.

4.5 What is *velocity head?*

4.6 What is *suction lift?*

4.7 What is *pumping water level?*

4.8 Water is pumped at 400 gpm to a water storage tank on the roof of a hospital. The gauge on the pump discharge line reads 90 psi. The difference in elevation between the gauge reading and the water level in the tank is 80 ft. What is the pressure loss (ft) in the piping?

4.9 A 10-hp pump delivers 1000 gpm. Assuming the pump is 100% efficient, what is the pressure (psi) against which the pump is operating?

4.10 Given the following information on a well, what is the drawdown when the pump is operating?

Static water level: 500 ft

Casing diameter: 12 inches

Well depth: 1200 ft

Pumping rate: 120 gpm

Pumping water level: 600 ft

Friction Head Loss

Materials or substances capable of flowing cannot flow absolutely freely. Nothing flows without encountering some type of resistance. Consider electricity, the flow of free electrons in a conductor. Whatever type of conductor used (i.e., copper, aluminum, silver, etc.) offers some resistance. In hydraulics, the flow of water/wastewater is analogous to the flow of electricity. Within a pipe or open channel, for instance, flowing water, like electron flow in a conductor, encounters resistance. However, resistance to the flow of water is generally termed friction loss or, more appropriately head loss.

TOPICS

Pipe and Open Flow Basics
Major Head Loss
Minor Head Loss

5.1 INTRODUCTION

The problem of water/wastewater flow in pipelines—the prediction of flow rate through pipes of given characteristics, the calculation of energy conversions therein, and so forth—is encountered in many applications of water/wastewater operations and practice. Although the subject of pipe flow embraces only those problems in which pipes flow completely full (as in water lines), also addressed in this chapter are pipes that flow partially full (wastewater lines, normally treated as open channels).

The solution of practical pipe flow problems (some of which have been presented earlier in the text) resulting from application of the energy principle, the equation of continuity, and the principle and equation of water resistance are also discussed. Resistance to flow in pipes is offered not only by long reaches of pipe but also by pipe fittings, such as bends and valves, which dissipate energy by producing relatively large-scale turbulence.

5.2 PIPE AND OPEN FLOW BASICS

To gain understanding of what friction head loss is all about, it is necessary to review a few terms presented earlier in the text and to introduce some new terms pertinent to the subject.[12]

🔵 *Laminar and Turbulent Flow*—**laminar flow** is ideal flow; that is, water particles moving along straight, parallel paths, in layers or streamlines. Moreover, in laminar flow, there is no turbulence in

---

[11] From Metcalf & Eddy, *Wastewater Engineering: Collection and Pumping of Wastewater.* New York: McGraw-Hill Book Company, p. 11, 1981.

[12] A more complete listing of hydraulic terms may be found in Lindeburg, M. R., *Civil Engineering Reference Manual,* 4th ed. San Carlos, CA: Professional Publications, Inc., pp. 5-2, 5-3, 1986.

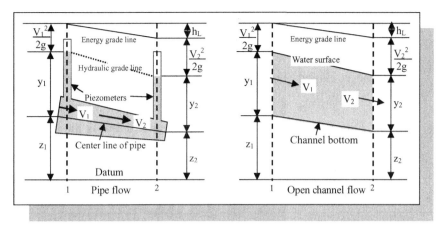

Figure 5.1
Comparison of pipe flow and open-channel flow. (Adapted from Metcalf & Eddy, 1981, p. 11.)

the water and no friction loss. This is not typical of normal pipe flow because the water velocity is too great, but is typical of groundwater flow.

Turbulent flow (characterized as "normal" for a typical water system) occurs when water particles move in a haphazard fashion and continually cross each other in all directions resulting in pressure losses along a length of pipe.

Hydraulic Grade Line (HGL)—recall from Chapter 3 that the hydraulic grade line (HGL), also shown in Figure 5.1, is a line connecting two points to which the liquid would rise at various places along any pipe or open channel if piezometers were inserted in the liquid. It is a measure of the pressure head available at these various points.

Note: *It is important to point out that when water flows in an open channel, the HGL coincides with the profile of the water surface.*

IMPORTANT

Energy Grade Line—the total energy of flow in any section with reference to some datum (i.e., a reference line, surface, or point) is the sum of the elevation head z, the pressure head y, and the velocity head $V^2/2g$. Figure 5.1 shows the **energy grade line** or **energy gradient,** which represents the energy from section to section. In the absence of frictional losses, the energy grade line remains horizontal, although the relative distribution of energy may vary between the elevation, pressure, and velocity heads. In all real systems, however, losses of energy occur because of resistance to flow, and the resulting energy grade line is **sloped** (i.e., the energy grade line is the slope of the specific energy line).

💧 *Specific Energy (E)*—sometimes called **specific head,** is the sum of the pressure head y and the velocity head $V^2/2g$. The specific energy concept is especially useful in analyzing flow in open channels.

💧 *Steady Flow*—occurs when the discharge or rate of flow at any cross section is constant.

💧 *Uniform and Non-uniform Flow*—**uniform flow** occurs when the depth, cross-sectional area, and other elements of flow are substantially constant from section to section. **Non-uniform flow** occurs when the slope, cross-sectional area, and velocity change from section to section. The flow through a Venturi section used for measuring flow is a good example.

💧 *Varied Flow*—flow in a channel is considered varied if the depth of flow changes along the length of the channel. The flow may be gradually varied or rapidly varied (i.e., when the depth of flow changes abruptly) as shown in Figure 5.2.

💧 *Slope* (gradient)—the head loss per foot of channel.

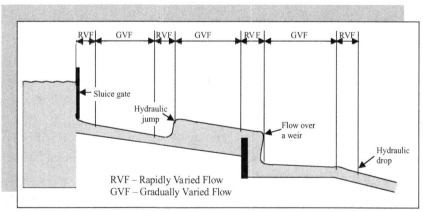

Figure 5.2
Varied flow.

5.3 MAJOR HEAD LOSS

Major head loss consists of pressure decreases along the length of pipe caused by friction created as water encounters the surfaces of the pipe. It typically accounts for most of the pressure drop in a pressurized or dynamic water system.

5.3.1 COMPONENTS OF MAJOR HEAD LOSS

In this section, we describe the components that contribute to major head loss: roughness, length, diameter, and velocity.

5.3.1.1 ROUGHNESS

Even when new, the interior surfaces of pipes are rough. The roughness varies, of course, depending on pipe material, corrosion (tuberculation and pitting), and age. Because normal flow in a water pipe is turbulent, the turbulence increases with pipe roughness, which, in turn, causes pressure to drop over the length of the pipe.

5.3.1.2 PIPE LENGTH

With every foot of pipe length, friction losses occur. The longer the pipe, the more head loss. Friction loss as a result of pipe length must be factored into head loss calculations.

5.3.1.3 PIPE DIAMETER

Generally, small diameter pipes have more head loss than large diameter pipes. This is the case because, in large diameter pipes, less of the water actually touches the interior surfaces of the pipe (encountering less friction) than in a small diameter pipe.

5.3.1.4 WATER VELOCITY

Turbulence in a water pipe is directly proportional to the speed (or velocity) of the flow. Thus, the velocity head also contributes to head loss.

Important Point: *For the same diameter pipe, when flow increases, head loss increases.*

IMPORTANT

5.3.2 CALCULATING MAJOR HEAD LOSS

The first practical equation used to determine pipe friction was developed in about 1850 by Darcy, Weisbach, and others. The equation or formula now known as the *Darcy-Weisbach* equation for circular pipes is

$$h_f = f\frac{LV^2}{D2g} \qquad (5.1)$$

In terms of the flow rate Q, the equation becomes

$$h_f = \frac{8fLQ^2}{\pi^2 g D^5} \qquad (5.2)$$

where

h_f = head loss, (ft)
f = coefficient of friction
L = length of pipe, (ft)
V = mean velocity, (ft/s)

D = diameter of pipe, (ft)
g = acceleration due to gravity, (32.2 ft/s^2)
Q = flow rate, (ft^3/s)

The Darcy-Weisbach formula as such was meant to apply to the flow of any fluid and into this friction factor was incorporated the degree of roughness and an element called the **Reynold's Number,** which was based on the viscosity of the fluid and the degree of turbulence of flow.

The Darcy-Weisbach formula is used primarily for determining head loss calculations in pipes. For making this determination in open channels, the *Manning Equation* was developed during the later part of the nineteenth century. Later, this equation was used for both open channels and closed conduits.

In the early 1900s, a more practical equation, the *Hazen-Williams* equation, was developed for use in making calculations related to water pipes and wastewater force mains:

$$Q = 0.435 \times CD^{2.63} \times S^{0.54} \qquad (5.3)$$

where

Q = flow rate, (ft^3/s)
C = coefficient of roughness (C decreases with roughness)
D = hydraulic radius r, (ft)
S = slope of energy grade line, (ft/ft)

5.3.2.1 C FACTOR

C factor, as used in the Hazen-Williams formula, designates the coefficient of roughness. C does not vary appreciably with velocity, and by comparing pipe types and ages, it includes only the concept of roughness, ignoring fluid viscosity and Reynold's Number.

Based on experience (experimentation), accepted tables of C factors have been established for pipe (see Table 5.1). Generally, C factor decreases by one with each year of pipe age. Flow for a newly designed system is often calculated with a C factor of 100, based on averaging it over the life of the pipe system.

TABLE 5.1. C Factors.*	
Type of Pipe	C Factor
Asbestos cement	140
Brass	140
Brick sewer	100
Cast iron	
10 years old	110
20 years old	90
Ductile iron, (cement lined)	140
Concrete or concrete lined	
Smooth, steel forms	140
Wooden forms	120
Rough	110
Copper	140
Fire hose (rubber lined)	135
Galvanized iron	120
Glass	140
Lead	130
Masonry conduit	130
Plastic	150
Steel	
Coal-tar enamel lined	150
New unlined	140
Riveted	110
Tin	130
Vitrified	120
Wood stave	120

*Adapted from Lindeburg, 1986, pp. 3–20.

Key Point: *A high C factor means a smooth pipe. A low C factor means a rough pipe.*

IMPORTANT

Note: *An alternate to calculating the Hazen-Williams formula, called an alignment chart, has become quite popular for fieldwork. The alignment chart can be used with reasonable accuracy.*

IMPORTANT

5.3.2.2 SLOPE

Slope is defined as the head loss per foot. In open channels, where the water flows by gravity, slope is the amount of incline of the pipe and is calculated as feet of drop per foot of pipe length (ft/ft). Slope is designed to be just enough to overcome frictional losses, so that the velocity remains constant, the water keeps flowing, and the solids will not settle in the conduit. In piped systems, where pressure loss for every foot of pipe is experienced, slope is not provided by slanting the pipe but instead by pressure added to overcome friction.

5.4 MINOR HEAD LOSS

In addition to the head loss caused by friction between the fluid and the pipe wall, losses also are caused by turbulence created by obstructions (i.e., valves and fittings of all types) in the line, changes in direction, and changes in flow area.

Note: *In practice, if minor head loss is less than 5% of the total head loss, it is usually ignored.*

IMPORTANT

REFERENCES

Lindeburg, M. R., *Civil Engineering Reference Manual,* 4th ed. San Carlos, CA: Professional Publications, Inc., 1986.

Metcalf & Eddy, *Wastewater Engineering: Collection and Pumping of Wastewater.* New York: McGraw-Hill Book Company, 1981.

Self-Test

5.1 What is *head loss?*

5.2 Explain *energy grade line.*

5.3 What is *steady flow?*

5.4 Define *slope.*

5.5 List four factors that affect head loss.

Basic Piping Hydraulics

Water, regardless of the source, is conveyed to the waterworks for treatment and is distributed to the users. Conveyance from the source to the point of treatment occurs by aqueducts, pipelines, or open channels, but the treated water is normally distributed in pressurized closed conduits. After use, whatever the purpose, the water becomes wastewater, which must be disposed of somehow, but almost always ends up being conveyed back to a treatment facility before being outfalled to some water body, to begin the cycle again.

We call this an urban water cycle, because it provides a human-generated imitation of the natural water cycle. Unlike the natural water cycle, however, without pipes, the cycle would be non-existent or, at the very least, short-circuited.

TOPICS

Piping Networks

Key Terms Used in This Chapter

EQUIVALENT PIPE THEORY	States that two pipes are equivalent if the head loss generated by the water velocity is the same in both pipes.
PARALLEL PIPE SYSTEMS	Two or more pipes of different ages, materials, sizes, or lengths, laid side by side, with the flow splitting among them.
SERIES PIPE SYSTEMS	Two or more pipes of different ages, materials, or sizes laid end to end.

6.1 INTRODUCTION

For use as water mains in a distribution system, pipes must be strong and durable in order to resist applied forces and corrosion. The pipe is subjected to internal pressure from the water and to external pressure from the weight of the backfill (soil) and vehicles above it. The pipe may also have to withstand water hammer. Damage due to corrosion or rusting may also occur internally because of the water quality or externally because of the nature of the soil conditions.

Pipes used in a wastewater system must be strong and durable to resist the abrasive and corrosive properties of the wastewater. Like water pipes, wastewater pipes must also be able to withstand stresses caused by the soil backfill material and the effect of vehicles passing above the pipeline.

Joints between wastewater collection/interceptor pipe sections should be flexible, but tight enough to prevent excessive leakage, either of sewage out of the pipe or groundwater into the pipe.

Of course, pipes must be constructed to withstand the expected conditions of exposure, and pipe configuration systems for water distribution and/or wastewater collection and interceptor systems must be properly designed and installed in terms of water hydraulics. Because the water/wastewater operator should have a basic knowledge of water hydraulics related to commonly used standard piping configurations, piping basics are briefly covered in this chapter.

6.2 PIPING NETWORKS

It would be far less costly and make for more efficient operation if municipal water and wastewater systems were built with separate single-pipe networks extending from treatment plant to user's residence or from user's sink or bathtub drain to the local wastewater treatment plant. Unfortunately, this ideal single-pipe scenario is not practical for real-world applications. Instead of a single piping system, a network of pipes is laid under the streets. Each of these piping networks is composed of different materials that vary (sometimes considerably) in diameter, length, and age. These networks range in complexity to varying degrees, and each of these joined-together pipes contributes energy losses to the system.

6.2.1 ENERGY LOSSES IN PIPE NETWORKS

Water/wastewater flow networks may consist of pipes arranged in series, parallel, or some complicated combination. In any case, an evaluation of friction losses for the flows is based on energy conservation principles applied to the flow junction points. Methods of computation depend on the particular piping configuration. In general, however, they involve establishing a sufficient number of simultaneous equations or employing a friction loss formula where the friction coefficient depends only on the roughness of the pipe [e.g., Hazen-Williams Equation—Equation (5.3)]. (Note: Demonstrating the procedure for making these complex computations is beyond the scope of this basic text. Only the "need to know" basic aspects of complex or compound piping systems are presented in this text.)

6.2.1.1 PIPES IN SERIES

When two pipes of different sizes or roughnesses are connected in series (see Figure 6.1), head loss for a given discharge, or discharge for a given head loss, may be calculated by applying the appropriate equation

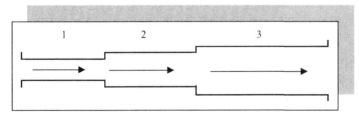

Figure 6.1
Pipes in series.

between the bonding points, taking into account all losses in the interval. Thus, head losses are cumulative.

Series pipes may be treated as a single pipe of constant diameter to simplify the calculation of friction losses. The approach involves determining an "equivalent length" of a constant diameter pipe that has the same friction loss and discharge characteristics as the actual series pipe system. In addition, application of the continuity equation to the solution allows the head loss to be expressed in terms of only one pipe size.

Note: *In addition to the head loss caused by friction between the water and the pipe wall, losses also are caused by minor losses: obstructions in the line, changes in directions, and changes in flow area. In practice, the method of equivalent length is often used to determine these losses. The method of equivalent length uses a table to convert each valve or fitting into an equivalent length of straight pipe.*

IMPORTANT

In making calculations involving pipes in series, remember these two important basic operational tenets:

1. The same flow passes through all pipes connected in series.

2. The total head loss is the sum of the head losses of all of the component pipes.

In some operations involving series networks where the flow is given and the total head loss is unknown, we can use the Hazen-Williams formula to solve for the slope and the head loss of each pipe as if they were separate pipes. Adding up the head losses to get the total head loss is then a simple matter.

Other series network calculations may not be as simple to solve using the Hazen-Williams Equation. For example, one problem is what diameter to use with varying sized pipes connected together in a series combination. Moreover, head loss is applied to both pipes (or other multiples), and it is not known how much loss originates from each one; thus, determining slope would be difficult—but not impossible.

In such cases, the **Equivalent Pipe Theory,** as mentioned earlier, can be used. Again, one single "Equivalent Pipe" is created that will carry the correct flow. This is practical because the head loss through it is the same as that in the actual system. The equivalent pipe can have any C factor and diameter, just as long as those same dimensions are maintained all the way through to the end. Keep in mind that the equivalent pipe must have the correct length, so that it will allow the correct flow through, which yields the correct head loss (the given head loss).[13]

6.2.1.2 PIPES IN PARALLEL

Two or more pipes connected (as in Figure 6.2) so that flow is first divided among the pipes and is then rejoined comprise a parallel pipe system. A parallel pipe system is a common method for increasing the capacity of an existing line. Determining flows in pipes arranged in parallel are also made by application of energy conservation principles—specifically, energy losses through all pipes connecting common junction points must be equal. Each leg of the parallel network is treated as a series piping system and is converted to a single equivalent length pipe. The friction losses through the equivalent length parallel pipes are then considered equal, and the respective flows are determined by proportional distribution.

Computations used to determine friction losses in parallel combinations may be accomplished using a simultaneous solution approach for a parallel system that has only two branches. However, if the parallel system

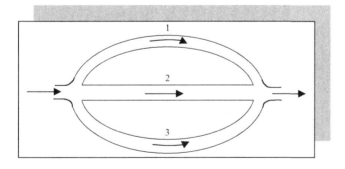

Figure 6.2
Pipe in parallel.

[13]For more information on how to use the equivalent pipe theory in making computations involving series or parallel pipe combinations, refer to Lindeburg, M. R., *Civil Engineering Reference Manual,* 4th ed. San Carbs, CA: Professional Publications, Inc., 1986.

has three or more branches, a modified procedure using the Hazen-Williams loss formula is easier.[14]

REFERENCE

Lindeburg, M. R., *Civil Engineering Reference Manual,* 4th ed. San Carlos, CA: Professional Publications, Inc., 1986.

[14]See Lindeburg, p. 3–26.

Self-Test

6.1 Define equivalent pipe theory.

6.2 Define parallel pipe systems.

6.3 Define series pipe systems.

Open-Channel Flow

Water is transported over long distances through aqueducts to locations where it is to be used and/or treated. Selection of an aqueduct type rests on such factors as topography, head availability, climate, construction practices, economics, and water quality protection. Along with pipes and tunnels, aqueducts may also include or be solely composed of open channels.[15]

TOPICS

Characteristics of Open-Channel Flow
Parameters Used in Open-Channel Flow
Open-Channel Flow Calculations
Open-Channel Flow: The Bottom Line

7.1 INTRODUCTION

Chapter 6 focused on flow in pressure conduits that are always full. In this chapter, water passage in open channels, which allows part of the water to be exposed to the atmosphere, is discussed. This type of channel—on open-flow channel—includes natural waterways, canals, culverts, flumes, and pipes flowing under the influence of gravity.

---

[15] Viessman, W., Jr. and Hammer, M. J., *Water Supply and Pollution Control,* 6th ed. Menlo Park, CA: Addison-Wesley, p. 119, 1998.

Key Terms Used in This Chapter

LAMINAR FLOW	Ideal water flow; water particles move along straight, parallel paths, in layers or streamlines. There is no turbulence in the water and no friction loss.
TURBULENT FLOW	Normal for a water system; water particles move in a haphazard fashion and continually cross each other in all directions.
UNIFORM FLOW	Occurs when the magnitude and direction of velocity do not change from point to point.
VARIED FLOW	Flow that has a changing depth along the water course, with respect to location, not time.
CRITICAL FLOW	Flow at the critical depth and velocity. Critical flow minimizes the specific energy and maximizes discharge.
SUBCRITICAL FLOW	Flow at greater than the critical depth (less than the critical velocity).
SUPERCRITICAL FLOW	Flow at less than the critical depth (greater than the critical velocity).

7.2 CHARACTERISTICS OF OPEN-CHANNEL FLOW[16]

Basic hydraulic principles apply in open-channel flow (with water depth constant), although there is no pressure to act as the driving force. Velocity head is the only natural energy this water possesses, and at normal water velocities, this is a small value ($V^2/2g$).

---

[16]Adapted from McGhee, T. J., *Water Supply and Sewerage,* 2nd ed. New York: McGraw-Hill, p. 45, 1991.

Several parameters can be (and often are) used to describe open-channel flow. However, we begin our discussion with a few characteristics including laminar or turbulent, uniform or varied, and subcritical, critical, or supercritical.

7.2.1 LAMINAR AND TURBULENT FLOW

Laminar and *turbulent* flow in open channels is analogous to that in closed pressurized conduits (i.e., pipes). It is important to point out, however, that flow in open channels is usually turbulent. In addition, there is no important circumstance in which laminar flow occurs in open channels in either water or wastewater unit processes or structures.

7.2.2 UNIFORM AND VARIED FLOW

Flow can be a function of time and location. If the flow quantity is invariant, it is said to be steady. *Uniform* flow is flow in which the depth, width, and velocity remain constant along a channel. That is, if the flow cross section does not depend on the location along the channel, the flow is said to be uniform. *Varied* or *non-uniform* flow involves a change in these, with a change in one producing a change in the others. Most circumstances of open-channel flow in water/wastewater systems involve varied flow. The concept of uniform flow is valuable, however, in that it defines a limit that the varied flow may be considered to be approaching in many cases.

 Note: *Uniform channel construction does not ensure uniform flow.*

IMPORTANT

7.2.3 CRITICAL FLOW

Critical flow (i.e., flow at the critical depth and velocity) defines a state of flow between two flow regimes. Critical flow coincides with minimum specific energy for a given discharge and maximum discharge for a given specific energy. Critical flow occurs in flow measurement devices at or near free discharges and establishes controls in open-channel flow. Critical flow occurs frequently in water/wastewater systems and is very important in their operation and design.

Important Point: *Critical flow minimizes the specific energy and maximizes discharge.*

IMPORTANT

7.3 PARAMETERS USED IN OPEN-CHANNEL FLOW

The three primary parameters used in open-channel flow are hydraulic radius, hydraulic depth, and slope, *S*.

7.3.1 HYDRAULIC RADIUS

The *hydraulic radius* is the ratio of area in flow to wetted perimeter.

$$r_H = \frac{A}{P} \qquad (7.1)$$

where

r_H = hydraulic radius
A = the cross-sectional area of the water
P = wetted perimeter

In open channels, it is of primary importance to maintain the proper velocity. This is the case, of course, because if velocity is not maintained, then flow stops (theoretically). To maintain velocity at a constant level, the channel slope must be adequate to overcome friction losses. As with other flows, calculation of head loss at a given flow is necessary, and the Hazen-Williams Equation is useful ($Q = 0.435 \times C \times d^{2.63} \times S^{0.54}$).

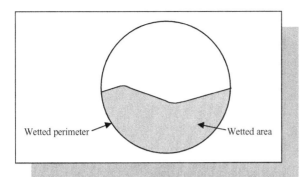

Wetted perimeter ──→ ──← Wetted area

Figure 7.1
Hydraulic radius.

Keep in mind that the concept of slope has not changed. The difference? The physical slope of a channel (ft/ft), equivalent to head loss, is now being measured, or calculated for.

The preceding seems logical, but there is a problem. The problem is with the diameter. In conduits that are not circular (grit chambers, contact basins, streams, and rivers), or in pipes only partially full (drains, wastewater gravity mains, sewers, etc.) where the cross-sectional area of the water is not circular, there is no diameter.

Because there is no diameter in a situation where the cross-sectional area of the water is not circular, we must use another parameter to designate the size of the cross section and the amount of it that contacts the sides of the conduit. This is where the **hydraulic radius** (r_H) comes in. The hydraulic radius is a measure of the efficiency with which the conduit can transmit water. Its value depends on pipe size and amount of fullness. Simply, we use the hydraulic radius to measure how much of the water is in contact with the sides of the channel or how much of the water is not in contact with the sides (see Figure 7.1).

IMPORTANT

Important Point: *For a circular channel flowing either full or half-full, the hydraulic radius is (D/4). Hydraulic radii of other channel shapes are easily calculated from the basic definition.*

7.3.2 HYDRAULIC DEPTH

The *hydraulic depth* is the ratio of area in flow to the width of the channel at the fluid surface. (Note that another name for hydraulic depth is the **hydraulic mean depth** or hydraulic radius.)

$$d_H = \frac{A}{w} \qquad (7.2)$$

where

d_H = hydraulic depth
A = area in flow
w = width of the channel at the fluid surface

7.3.3 SLOPE, S

The *slope, S,* in open-channel equations is the slope of the energy line. If the flow is uniform, the slope of the energy line will parallel the water surface and channel bottom. In general, the slope can be calculated from the Bernoulli equation as the energy loss per unit length of channel.

$$S = \frac{dH}{dL} \qquad (7.3)$$

S = slope
dH = total hydraulic head
dL = channel length

7.4 OPEN-CHANNEL FLOW CALCULATIONS

It was stated earlier that the calculation for head loss at a given flow is typically accomplished by using the Hazen-Williams Equation. In open-channel flow, although the concept of slope has not changed, the problem arises with the diameter. Again, in pipes only partially full where the cross-sectional area of the water is not circular, there is no diameter. Thus, the hydraulic radius is used for these non-circular areas.

In the original version of the Hazen-Williams Equation, the hydraulic radius was incorporated. Moreover, similar versions developed by Chezy and Manning and others incorporated the hydraulic radius. For use in open channels, **Manning's Formula** has become most commonly used:

$$Q = \frac{1.5}{n} A \times R^{66} \times S^5 \qquad (7.4)$$

where

Q = channel discharge capacity (ft^3/s)
1.5 = constant
n = channel roughness coefficient
A = cross-sectional flow area, (ft^2)
R = hydraulic radius of the channel, (ft)
S = slope of the channel bottom, dimensionless

TABLE 7.1. Manning Roughness Coefficient, n.			
Type of Conduit	n	Type of Conduit	n
Pipe			
Cast iron, coated	0.012–0.014	Cast iron, uncoated	0.013–0.015
Wrought iron, galvanized	0.015–0.017	Wrought iron, black	0.012–0.015
Steel, riveted and spiral	0.015–0.017	Corrugated	0.021–0.026
Wood stave	0.012–0.013	Cement surface	0.010–0.013
Concrete	0.012–0.017	Vitrified	0.013–0.015
Clay, drainage tile	0.012–0.014		
Lined Channels			
Metal, smooth semicircular	0.011–0.015	Metal, corrugated	0.023–0.025
Wood, planed	0.010–0.015	Wood, unplaned	0.011–0.015
Cement lined	0.010–0.013	Concrete	0.014–0.016
Cement rubble	0.017–0.030	Grass	–0.020
Unlined Channels			
Earth, straight and uniform	0.017–0.025	Earth, dredged	0.025–0.033
Earth, winding	0.023–0.030	Earth, stony	0.025–0.040
Rock, smooth and uniform	0.025–0.035	Rock, jagged and irregular	0.035–0.045

The hydraulic radius of a channel is defined as the ratio of the flow area to the wetted perimeter, P. In formula form, $R = A/P$. The new component is n (the roughness coefficient) and depends on the material and age for a pipe or lined channel and on topographic features for a natural streambed. It approximates roughness in open channels and can range from a value of 0.01 for a smooth clay pipe to 0.1 for a small natural stream. The value of n commonly assumed for concrete pipes or lined channels is 0.013, and n values decrease as the channels get smoother (see Table 7.1).

The following example illustrates the application of Manning's Formula for a channel with a rectangular cross section.

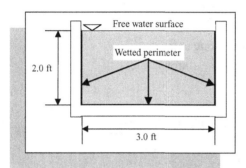

Figure 7.2
For Example 7.1.

EXAMPLE 7.1

Problem: A rectangular drainage channel is 3 ft wide and is lined with concrete, as illustrated in Figure 7.2. The bottom of the channel drops in elevation at a rate of 0.5 ft. per 100 ft. What is the discharge in the channel when the depth of water is 2 ft?

Solution:
Assume $n = 0.013$.

Referring to Figure 7.2, the cross-sectional flow area $A = 3$ ft $\times$ 2 ft $= 6$ ft^2, and the wetted perimeter $P = 2$ ft $+ 3$ ft $+ 2$ ft $= 7$ ft. The hydraulic radius $R = A/P = 6$ ft^2/7 ft $= 0.86$ ft. The slope $S = 0.5/100 = 0.005$.

Applying Manning's Formula:

$$Q = \frac{2.0}{0.013} \times 6 \times 0.86^{0.66} \times 0.005^{0.5}$$
$$Q = 59 \ cfs$$

7.5 OPEN-CHANNEL FLOW: THE BOTTOM LINE

It has been stated that when water flows in a pipe or channel with a **free surface** exposed to the atmosphere, it is called *open-channel flow.* We also know that gravity provides the motive force, the constant push, while friction resists the motion and causes energy expenditure. River and stream flow is open-channel flow. Flow in sanitary sewers and storm water drains is also open-channel flow, except in force mains where the water is pumped under pressure.

The key to solving storm water and/or sanitary sewer routine problems is a condition known as **steady uniform flow;** that is, we assume steady uniform flow. Steady flow, of course, means that the discharge is constant with time. Uniform flow means that the slope of the water surface and the cross-sectional flow area are also constant. It is common practice to call a length of channel, pipeline, or stream that has a relatively constant slope and cross section a *reach.*[17]

The slope of the water surface, under steady uniform flow conditions, is the same as the slope of the channel bottom. The hydraulic grade line (HGL) lies along the water surface, and, as in pressure flow in pipes, the HGL slopes downward in the direction of flow. Energy loss is evident as

[17] See Nathanson, J. A., *Basic Environmental Technology: Water Supply, Waste Management, and Pollution Control,* 2nd ed. Upper Saddle River, NJ: Prentice Hall, p. 34, 1997.

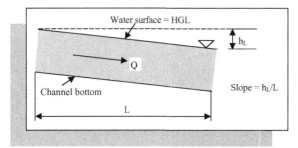

Figure 7.3
Steady uniform open-channel flow—where the slope of the water surface (or HGL) is equal to the slope of the channel bottom.

the water surface elevation drops. Figure 7.3 illustrates a typical profile view of uniform steady flow. The slope of the water surface represents the rate of energy loss.

Note: *Rate of energy loss (see Figure 7.3) may be expressed as the ratio of the drop in elevation of the surface in the reach to the* IMPORTANT *length of the reach.*

Figure 7.4 shows typical cross sections of open-channel flow. In Figure 7.4(a), the pipe is only partially filled with water, and there is a free surface at atmospheric pressure. This is still open-channel flow, although the pipe is a closed underground conduit. Remember, the important point is that gravity, and not a pump, is moving the water.

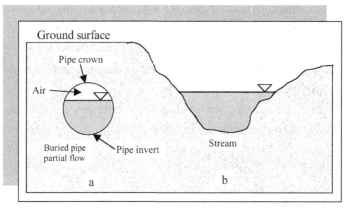

Figure 7.4
Shows open-channel flow, whether in a surface stream or in an underground pipe. (Adaptation from Nathanson, 1997, p. 35).

REFERENCES

McGhee, T. J., *Water Supply and Sewerage,* 2nd ed. New York: McGraw-Hill, 1991.

Nathanson, J. A., *Basic Environmental Technology: Water Supply, Waste Management, and Pollution Control,* 2nd ed. Upper Saddle River, NJ: Prentice Hall, 1997.

Viessman, W., Jr. and Hammer, M. J., *Water Supply and Pollution Control,* 6th ed. Menlo Park, CA: Addison-Wesley, 1998.

Self-Test

7.1 Distinguish between critical flow, subcritical flow, and supercritical flow.

7.2 Define hydraulic radius.

7.3 Define hydraulic depth.

7.4 Critical flow minimizes the specific _____ and _____ discharge.

7.5 Flow less than the critical velocity is known as _____.

Flow
Measurement

While it is clear that maintaining water/wastewater flow is at the heart of any treatment process, clearly, it is the measurement of flow that is essential to ensuring the proper operation of a water/wastewater treatment system. Few knowledgeable operators would argue with this statement. Hauser (1996) asks: "Why measure flow?" Then she explains: "The most vital activities in the operation of water and wastewater treatment plants are dependent on a knowledge of how much water is being processed."[18]

TOPICS

Flow Measurement: The Old-Fashioned Way
Basis of Traditional Flow Measurement
Flow Measuring Devices

8.1 INTRODUCTION

In the chapter opening, Hauser makes clear that flow measurement is not only important, but also routine, in water/wastewater operations. Hauser also points out that there are several reasons to measure flow in a treatment plant. The American Water Works Association[19] lists several additional reasons to measure flow:

💧 The flow rate through the treatment processes needs to be controlled so that it matches distribution system use.

💧 It is important to determine the proper feed rate of chemicals added in the processes.

[18]Adapted from Hauser, B. A., *Practical Hydraulics Handbook,* 2nd ed. Boca Raton, FL: Lewis Publishers, p. 91, 1996.
[19]From *Water Treatment: Principles and Practices of Water Supply Operations,* 2nd ed. Denver, CO: American Water Works Association, pp. 449–450, 1995.

Key Terms Used in This Chapter

FLOW	The quantity of water that passes a point in a given unit of time.
ORIFICE FLOWMETER	A flat plate across the line of flow that passes through a small hole in its center; a pressure differential meter.
OPEN-CHANNEL FLOW	Gravity flow in a pipe or open conduit with a free surface at atmospheric pressure.
VENTURI FLOWMETER	A section of constricted pipe through which the flow passes; a pressure differential meter.
PARSHALL FLUME	A constricted section in an open channel, for measuring flow rate.
MAGNETIC FLOWMETER	A flowmeter using a magnetic field to induce a voltage that is proportional to water velocity and convertible to flow.
ULTRASONIC FLOWMETER	Uses ultrasound frequency differential to detect water velocity or depth, which is then convertible to flow.
TURBINE FLOWMETER	Flow passes through a chamber enclosing a rotor that turns in response to velocity.
POSITIVE DISPLACEMENT FLOWMETER	The measuring chamber encloses a disc or piston that oscillates to pass the flow and transmits the movement to a register, which records.
WEIR	A flat obstruction placed across the line of flow in an open channel and over which the flow passes; reading is by depth measurement.

● The detention times through the treatment processes must be calculated. This is particularly applicable to surface water plants that must meet $C \times T$ values required by the Surface Water Treatment Rule.

● Flow measurement allows operators to maintain a record of water furnished to the distribution system for periodic comparison with the total water metered to customers. This provides a measure of "water accounted for" or, conversely (as pointed out earlier by Hauser), the amount of water wasted, leaked, or otherwise not paid for; that is, lost water.

● Flow measurement allows operators to determine the efficiency of pumps. (Note: Pumps are covered in greater detail in Chapter 9). Pumps that are not delivering their designed flow rate are probably not operating at maximum efficiency, and so power is being wasted.

● For well systems, it is very important to maintain records of the volume of water pumped and the hours of operation for each well. The periodic computation of well pumping rates can identify problems such as worn pump impellers and blocked well screens.

● Reports that must be furnished to the state by most water systems include records of raw and finished water pumpage.

● Wastewater generated by a treatment system must also be measured and recorded.

● Individual meters are often required for the proper operation of individual pieces of equipment. For example, the makeup water to a fluoride saturator is always metered to assist in tracking the fluoride feed rate.

Simply put, *measurement of flow is essential for operation, process control, and recordkeeping of water and wastewater*
IMPORTANT *treatment plants.*

All of the uses just discussed create the need, obviously, for a number of flow-measuring devices, often with different capabilities. In this chapter, we discuss many of the major flow measuring devices currently used in water/wastewater operations.

8.2 FLOW MEASUREMENT: THE OLD-FASHIONED WAY

An approximate, but very simple, method to determine open-channel flow has been used for many years. The procedure involves measuring the velocity of a floating object moving in a straight uniform reach of the channel or stream. If the cross-sectional dimensions of the channel are known and the depth of flow is measured, then flow area can be computed. From the relationship $Q = A \times V$, the discharge Q can be estimated.

In preliminary fieldwork, this simple procedure is useful in obtaining a ballpark estimate for the flow rate, but is not suitable for routine measurements.

EXAMPLE 8.1

Problem: A floating object is placed on the surface of water flowing in a drainage ditch and is observed to travel a distance of 20 meters downstream in 30 seconds. The ditch is 2 meters wide, and the average depth of flow is estimated to be 0.5 meters. Estimate the discharge under these conditions.

Solution: The flow velocity is computed as distance over time, or

$$V = D/T = 20 \text{ m}/30 \text{ s} = 0.67 \text{ m/s}$$

The channel area is $A = 2 \text{ m} \times 0.5 \text{ m} = 1.0 \text{ m}^2$
The discharge $Q = A \times V = 1.0 \text{ m}^2 \times 0.67 \text{ m/s} = 0.67 \text{ m}^3/\text{s}$

8.3 BASIS OF TRADITIONAL FLOW MEASUREMENT

Flow measurement can be based on flow rate, or flow amount. *Flow rate* is measured in gallons per minute (gpm), million gallons per day (MGD), or cubic feet per second (cfs). Water/wastewater operations need flow rate meters to determine process variables within the treatment plant, in wastewater collection, and in potable water distribution. Typically, flow rate meters used are pressure differential meters, magnetic meters, and ultrasonic meters. Flow rate meters are designed for metering flow in closed pipe or open-channel flow.

Flow amount is measured in either gallons (gal) or in cubic feet (cu ft). Typically, a totalizer, which sums up the gallons or cubic feet that pass through the meter, is used. Most service meters are of this type. They are used in private, commercial, and industrial activities where the total amount of flow measured is used in determining customer billing. In wastewater treatment, where sampling operations are important, automatic composite sampling units—flow proportioned to grab a sample every so many gallons—are used. Totalizer meters can be the velocity (propeller or turbine), positive displacement, or compound types. In addition, weirs and flumes are used extensively for measuring flow in wastewater treatment plants because they are not affected (to a degree) by dirty water or floating solids.

In the sections that follow, each of these measuring devices is briefly discussed.

8.4 FLOW-MEASURING DEVICES

In recent decades, flow-measurement technology has evolved rapidly from the "old fashioned way" of measuring flow discussed in Section 8.2 to the use of simple practical measuring devices to much more sophisticated devices. Physical phenomena discovered centuries ago have been the starting point for many of the viable flowmeter designs used today. Moreover, the recent technology explosion has enabled flowmeters to handle many more applications than could have been imagined centuries ago.

8.4.1 DIFFERENTIAL PRESSURE FLOWMETERS[20]

For many years, *differential pressure* flowmeters have been the most widely applied flow-measuring device for water flow in pipes that require accurate measurement at reasonable cost. The differential pressure type of flowmeter makes up the largest segment of the total flow measurement devices currently being used.

This type of device has a flow restriction in the line that causes a differential pressure or "head" to be developed between the two measurement locations. Differential pressure flowmeters are also known as head meters, and, of all the head meters, the orifice flowmeter is the most widely applied device.

The advantages of differential pressure flowmeters include

- simple construction

- relatively inexpensive

- no moving parts

- external transmitting instruments

- low maintenance

- wide application of flowing fluid

- ease of instrument and range selection

- extensive product experience and performance database

  Disadvantages include

- flow rate is a nonlinear function of the differential pressure

- low flow rate rangeability with normal instrumentation

[20]Adapted from Husain, Z. D., and Sergesketter, M. J., Differential Pressure Flowmeters, in *Flow Measurement,* Spitzer, D. W. (ed.). Research Triangle Park, NC: Instrument Society of America, pp. 119–160, 1991.

8.4.1.1 OPERATING PRINCIPLE

Differential pressure flowmeters operate on the principle of measuring pressure at two points in the flow, which provides an indication of the rate of flow that is passing by. The difference in pressures between the two measurement locations of the flowmeter is the result of the change in flow velocities. Simply, there is a set relationship between the flow rate and volume, so the meter instrumentation automatically translates the differential pressure into a volume of flow. The volume of flow rate through the cross-sectional area is given by,

$$Q = A \times v \, (average)$$

where:

Q = the volumetric flow rate
A = flow in the cross-sectional area
v = the average fluid velocity

8.4.1.2 TYPES OF DIFFERENTIAL PRESSURE FLOWMETERS

Differential pressure flowmeters operate on the principle of developing a differential pressure across a restriction that can be related to the fluid flow rate.

Note: *Optimum measurement accuracy is maintained when the flowmeter is calibrated, the flowmeter is installed in accordance* IMPORTANT *with standards and codes of practice, and the transmitting instruments are periodically calibrated.*

The most commonly used differential pressure flowmeter types used in water/wastewater treatment are

1. orifice

2. Venturi

3. nozzle

4. Pitot-static tube

8.4.1.2.1 ORIFICE

The most commonly applied *orifice* is a thin, **concentric,** and flat metal plate with an opening in the plate (see Figure 8.1), installed perpendicular to the flowing stream in a circular conduit or pipe. Typically, a sharp-edged hole is bored in the center of the orifice plate. As the flowing water passes through the orifice, the restriction causes an increase in velocity. A concurrent decrease in pressure occurs as potential energy (static pressure) is converted into kinetic energy (velocity). As the water leaves the orifice, its velocity decreases, and its pressure increases as kinetic energy is converted back into potential energy according to the laws of conservation of energy. However, there is always some permanent pressure loss due to friction, and the loss is a function of the ratio of the diameter of the orifice bore (d) to the pipe diameter (D).

For dirty water applications (i.e., wastewater), a concentric orifice plate will eventually have impaired performance due to dirt buildup at the plate. Instead, **eccentric** or **segmental** orifice plates (see Figure 8.2) are often used. Measurements are typically less accurate than those obtained from the concentric orifice plate. Eccentric or segmental orifices are rarely applied in current practice.

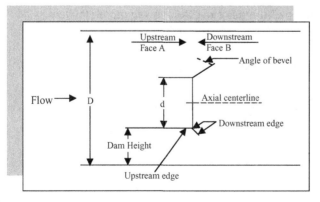

Figure 8.1
Orifice plate.

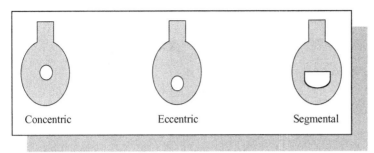

Figure 8.2
Types of Orifice plates.

The orifice differential pressure flowmeter is the lowest cost differential flowmeter, is easy to install, and has no moving parts. However, it also has high permanent head loss and higher pumping costs, and accuracy is affected with wear or damage.

Important Points: *Orifice meters are not recommended for permanent installation to measure wastewater flows; solids in the water easily catch on the orifice, throwing off accuracy. For installation, it is necessary to have 10 diameters of straight pipe ahead of the orifice meter to create a smooth flow pattern and five diameters of straight pipe on the discharge side.*

8.4.1.2.2 VENTURI

A *Venturi* is a restriction with a relatively long passage with smooth entry and exit (see Figure 8.3). It has long life expectancy, simplicity of construction, and relatively high pressure recovery (i.e., produces less permanent pressure loss than a similar sized orifice), but is more expensive, is not linear with flow rate, and is the largest and heaviest differential pressure flowmeter. It is often used in wastewater flows because the smooth entry allows foreign material to be swept through instead of building up as it would in front of an orifice.

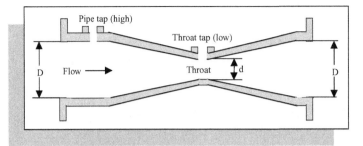

Figure 8.3
Venturi tube.

8.4.1.2.3 NOZZLE

Flow nozzles have a smooth entry and sharp exit (see Figure 8.4). For the same differential pressure, the permanent pressure loss of a nozzle is of the same order as that of an orifice, but it can handle wastewater and abrasive fluids better than an orifice can. Note that for the same line size and flow rate, the differential pressure at the nozzle is lower than the differential pressure for an orifice; hence, the total pressure loss is lower than that of an orifice. Nozzles are primarily used in steam service because of their rigidity, which makes them dimensionally more stable at high temperatures and velocities than orifices.

IMPORTANT **Note:** *A useful characteristic of nozzles is that they reach a critical flow condition, that is, a point at which further reduction in downstream pressure does not produce a greater velocity through the nozzle. When operated in this mode, nozzles are very predictable and repeatable.*

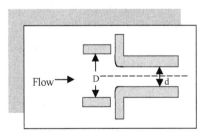

Figure 8.4
Long radius flow nozzle.

8.4.1.2.4 PITOT TUBE

A *Pitot tube* is a point velocity measuring device (see Figure 8.5). It has an impact port; as fluid hits the port, its velocity is reduced to zero, and kinetic energy (velocity) is converted to potential energy (pressure head). The pressure at the impact port is the sum of the static pressure and the velocity head. The pressure at the impact port is also known as stagnation pressure or total pressure. The pressure difference between the impact pressure and the static pressure measured at the same point is the velocity head. The flow rate is the product of the measured velocity and the cross-sectional area at the point of measurement. Note that the Pitot tube has negligible permanent pressure drop in the line, but the impact port must be located in the pipe where the measured velocity is equal to the average velocity of the flowing water through the cross section.

8.4.2 MAGNETIC FLOWMETERS[21]

Magnetic flowmeters are relatively new to the water/wastewater industry. They are volumetric flow devices designed to measure the flow of electrically conductive liquids in a closed pipe. They measure the flow

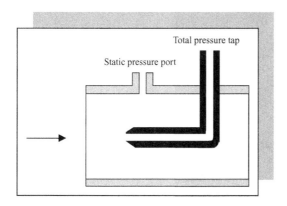

Figure 8.5
Pitot tube.

[21]Adapted from *Flow Instrumentation: A Practical Workshop on Making Them Work.* Sacramento, CA: The Water & Wastewater Instrumentation Testing Association and United States Environmental Protection Agency, Section A, May 16–17, 1991.

rate based on the voltage created between two electrodes (in accordance with Faraday's Law of Electromagnetic Induction) as the water passes through an electromagnetic field (see Figure 8.6). Induced voltage is proportional to flow rate. Voltage depends on magnetic field strength (constant), distance between electrodes (constant), and velocity of flowing water (variable).

Properties of the magnetic flowmeter include (1) minimal head loss (no obstruction with line size meter); (2) no effect on flow profile; (3) suitable for size range between 0.1 inch and 120 inches; (4) an accuracy rating of 0.5–2% of flow rate; and (5) measures forward or reverse flow.

The advantages of magnetic flowmeters include

- obstructionless flow

- minimal head loss

- wide range of sizes

- bi-directional flow measurement

- variations in density, viscosity, pressure, temperature yield negligible effect

- can be used for wastewater

- no moving parts

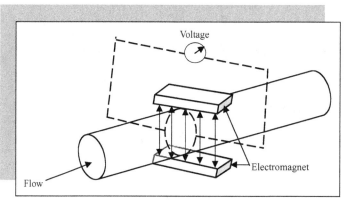

Figure 8.6
Magnetic flowmeter.

Disadvantages include

- metered liquid must be conductive (but this type of meter would not be used on clean fluids anyway)

- bulky, expensive in smaller sizes

The combination of the magnetic flowmeter and the transmitter is considered as a system. A typical system, schematically illustrated in Figure 8.7, shows a transmitter mounted remote from the magnetic flowmeter. Some systems are available with transmitters mounted integral to the magnetic flowmeter. Each device is individually calibrated during the manufacturing process, and the accuracy statement of the magnetic flowmeter includes both pieces of equipment. One is not sold or used without the other.

It is also interesting to note that since 1983 almost every manufacturer now offers the microprocessor-based transmitter.

Regarding minimum piping straight run requirements, magnetic flowmeters are quite forgiving of piping configuration. The downstream side of the magnetic flowmeter is much less critical than the upstream side. Essentially, all that is required of the downstream side is that sufficient backpressure is provided to keep the magnetic flowmeter full of liquid during flow measurement. Two diameters downstream should be acceptable.[22]

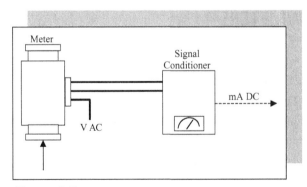

Figure 8.7
Magnetic flowmeter system.

[22]See Mills, R. C., Magnetic Flowmeters. In *Flow Measurement,* Spitzer, D. W. (ed.). Research Triangle Park, NC: Instrument Society of America, pp. 175–219, 1991.

Note: *Magnetic flowmeters are designed to measure conductive liquids only. If air or gas is mixed with the liquid, the output* IMPORTANT *becomes unpredictable.*

8.4.3 ULTRASONIC FLOWMETERS

Ultrasonic flowmeters use an electronic transducer to send a beam of ultrasonic sound waves through the water to another transducer on the opposite side of the unit. The velocity of the sound beam varies with the liquid flow rate, so the beam can be electronically translated to indicate flow volume.

Two types of ultrasonic flowmeters are generally used for closed pipe flow measurements. The first (time of flight or transit time) usually uses pulse transmission and is for clean liquids, while the second (Doppler) usually uses continuous wave transmission and is for dirty liquids.

8.4.3.1 TIME-OF-FLIGHT ULTRASONIC FLOWMETERS[23]

Time-of-flight flowmeters make use of the difference in the time for a sonic pulse to travel a fixed distance, first in the direction of flow and then against the flow. This is accomplished by opposing transceivers positioned on diagonal paths across meter spool as shown in Figure 8.8. Each transmits and receives ultrasonic pulses with flow and against flow. The fluid velocity is directly proportional to time difference of pulse travel.

---

[23] Adapted from Brown, A. E., Ultrasonic Flowmeters, in *Flow Measurement,* Spitzer, D. W. (ed.). Research Triangle Park, NC: Instrument Society of America, pp. 415–432, 1991.

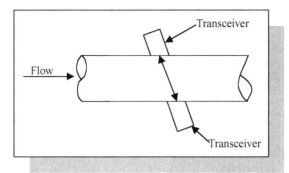

Figure 8.8
Time-of-flight ultrasonic flowmeter.

The time-of-flight ultrasonic flowmeter operates with minimal head loss and has an accuracy range of 1% to 2.5% full scale; they can be mounted as integral spool piece transducers or as externally mountable clamp-ons. They can measure flow accurately when properly installed and applied.

The advantages of time-of-flight ultrasonic flowmeters include

● no obstruction to flow

● minimal head loss

● clamp-ons

  • can be portable

  • no interruption of flow

● no moving parts

● linear over wide range

● wide range of pipe sizes

● bi-directional flow measurement

Disadvantages include

● sensitive to solids or bubble content

  • interfere with sound pulses

- sensitive to flow disturbances

- alignment of transducers is critical

- clamp-on—pipe walls must freely pass ultrasonic pulses

8.4.3.2 DOPPLER TYPE ULTRASONIC FLOWMETERS

Doppler ultrasonic flowmeters make use of the Doppler frequency shift caused by sound scattered or reflected from moving particles in the flow path. Doppler meters are not considered to be as accurate as time-of-flight flowmeters. However, they are very convenient to use and generally are more popular and less expensive than time-of-flight flowmeters.

In operation, a propagated ultrasonic beam is interrupted by particles in moving fluid and is reflected toward a receiver. The difference of propagated and reflected frequencies is directly proportional to fluid flow rate.

Ultrasonic Doppler flowmeters feature minimal head loss with an accuracy of 2% to 5% full scale. They are either integral spool piece transducer type or externally mountable clamp-ons.

The advantages of the Doppler ultrasonic flowmeter include

- no obstruction to flow

- minimal head loss

- clamp-on

  - can be portable

  - no interruption of flow

- no moving parts

- linear over wide range

- wide range of pipe sizes

- low installation and operating costs

- bi-directional flow measurement

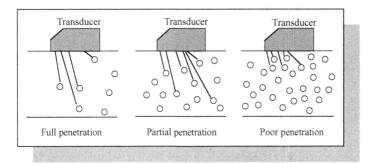

Figure 8.9
Particle concentration effect. The more particles there are, the more error.

The disadvantages include

● requires minimum concentration and size of solids or bubbles for reliable operation (see Figure 8.9)

● requires a minimum speed to maintain suspension

● clamp-on type limited to sonically conductive pipe

8.4.4 VELOCITY FLOWMETERS[24]

Velocity or *turbine* flowmeters use a propeller or turbine to measure the velocity of the flow passing the device (see Figure 8.10). The velocity is then translated into a volumetric amount by the meter register. Sizes exist from a variety of manufacturers to cover the flow range from 0.001 gpm to more than 25,000 gpm for liquid service. End connections are available to meet the various piping systems. The flowmeters are typically manufactured of stainless steel but are also available in a wide variety of materials, including plastic. Velocity meters are applicable to all clean fluids and are particularly well suited for measuring intermediate flow rates on clean water.

[24] Adapted from Oliver, P. D., Turbine Flowmeters, in *Flow Measurement,* Spitzer, D. W. (ed.). Research Triangle Park, NC: Instrument Society of America, pp. 373–414, 1991.

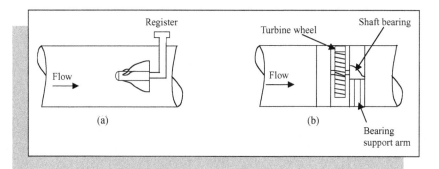

Figure 8.10
(a) Propeller meter, (b) turbine meter.

The advantages of the velocity meter include

● high accuracy

● corrosion-resistant materials

● long-term stability

● liquid or gas operation

● wide operating range

● low pressure drop

● wide temperature and pressure limits

● high shock capability

● wide variety of electronics available

As shown in Figure 8.10, a turbine flowmeter consists of a rotor mounted on a bearing and shaft in a housing. The fluid to be measured is passed through the housing, causing the rotor to spin with a rotational speed proportional to the velocity of the flowing fluid within the meter. A device to measure the speed of the rotor is employed to make the actual flow measurement. The sensor can be a mechanically gear-driven shaft to a meter or an electronic sensor that detects the passage of each rotor blade generating a pulse. The rotational speed of the sensor shaft and the frequency of the pulse are proportional to the volumetric flow rate through the meter.

8.4.5 POSITIVE-DISPLACEMENT FLOWMETERS[25]

Positive-displacement flowmeters are most commonly used for customer metering; they have long been used to measure liquid products. These meters are very reliable and accurate for low flow rates because they measure the exact quantity of water passing through them. Positive-displacement flowmeters are frequently used for measuring small flows in a treatment plant because of their accuracy. Repair or replacement is easy because they are so common in the distribution system.

In essence, a positive-displacement flowmeter is a hydraulic motor with high volumetric efficiency that absorbs a small amount of energy from the flowing stream. This energy is used to overcome internal friction in driving the flowmeter and its accessories and is reflected as a pressure drop across the flowmeter. Pressure drop is regarded as a necessary evil that must be minimized. It is the pressure drop across the internals of a positive-displacement flowmeter that actually creates a hydraulically unbalanced rotor, which causes rotation.

Simply, a positive-displacement flowmeter is one that continuously divides the flowing stream into known volumetric segments, isolates the segments momentarily, and returns them to the flowing stream while counting the number of displacements.

A positive-displacement flowmeter can be broken down into three basic components: the external housing, the measuring unit, and the counter drive train.

The external housing is the pressure vessel that contains the product being measured.

The measuring unit is a precision metering element and is made up of the measuring chamber and the displacement mechanism. The most common displacement mechanisms include the oscillating piston, sliding vane, oval gear, trirotor, birotor, and nutating disc types (see Figure 8.11).

[25]Barnes, R. G., Positive Displacement Flowmeters for Liquid Measurement, in *Flow Measurement,* Spitzer, D. W. (ed.). Research Triangle Park, NC: Instrument Society of America, pp. 315–322, 1991.

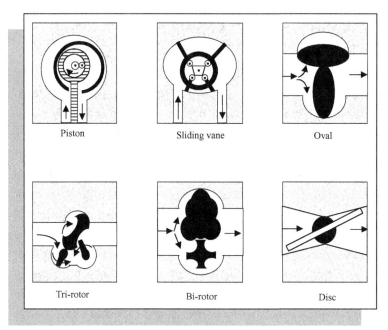

Figure 8.11
Six common positive-displacement meter principles.

The counter drive train is used to transmit the internal motion of the measuring unit into a usable output signal. Many positive-displacement flowmeters use a mechanical gear train that requires a rotary shaft seal or packing gland where the shaft penetrates the external housing.

The positive-displacement flowmeter can offer excellent accuracy, repeatability, and reliability in many applications.

The positive-displacement flowmeter has satisfied many needs in the past and should play a vital role in serving the future needs as required.

8.4.6 OPEN-CHANNEL FLOW MEASUREMENT[26]

The majority of industrial liquid flows are carried in closed conduits that flow completely full and under pressure. However, this is not the case

[26] Adapted from Grant, D. M., Open Channel Flow Measurement, in *Flow Measurement.* Research Triangle Park, NC: Instrument Society of America, pp. 252–290, 1991.

for high volume flows of liquids in waterworks, sanitary, and stormwater systems that are commonly carried in open channels. Flow in open channels is characterized by low system heads and high volumetric flow rates.

The most commonly used method of measuring the rate of flow in open-channel flow configurations is that of **hydraulic structures.** In this method, flow in an open channel is measured by inserting a hydraulic structure into the channel, which changes the level of liquid in or near the structure. By selecting the shape and dimensions of the hydraulic structure, the rate of flow through or over the restriction will be related to the liquid level in a known manner. Thus, the flow rate through the open channel can be derived from a single measurement of the liquid level in or near the structure.

The hydraulic structures used in measuring flow in open channels are known as primary measuring devices and may be divided into two broad categories—weirs and flumes—which are covered in the following subsections.

8.4.6.1 WEIRS

The *weir* is a widely used device to measure open-channel flow. As can be seen in Figure 8.12, a weir is simply a dam or obstruction placed in the channel so that water backs up behind it and then flows over it. The sharp crest or edge allows the water to spring clear of the weir plate and to fall freely in the form of a **nappe.**

As Nathanson[27] points out, when the nappe discharges freely into the air, there is a hydraulic relationship between the height or depth of water flowing over the weir crest and the flow rate. This height, the vertical distance between the crest and the water surface, is called the **head on the weir;** it can be measured directly with a meter or yardstick or automatically by float-operated recording devices. Two common weirs, rectangular and triangular, are shown in Figure 8.13.

---

[27]Nathanson, J. A., *Basic Environmental Technology: Water Supply, Waste Management, and Pollution Control,* 2nd ed. Upper Saddle River, NJ: Prentice Hall, p. 39, 1997.

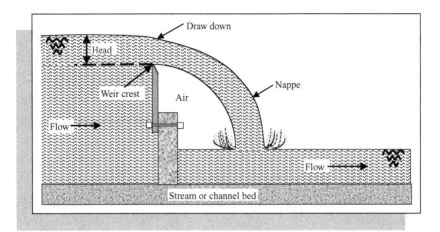

Figure 8.12
Side view of a weir.

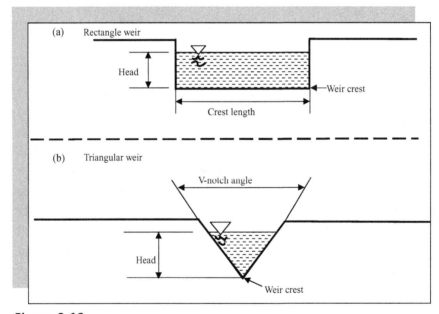

Figure 8.13
(a) Rectangular weir, (b) triangular V-notch weir.

Rectangular weirs are commonly used for large flows [see Figure 8.13(a)]. The formula used to make rectangular weir computations is

$$Q = 3.33 \times L \times h^{1.5} \qquad (8.1)$$

where

Q = flow
L = width of weir
h = head on weir (measured from edge of weir in contact with the water, up to the water surface)

EXAMPLE 8.2

Problem: A weir 4-ft high extends 15 ft across a rectangular channel in which there are 80 cfs flowing. What is the depth just upstream from the weir?

Solution:

$Q = 3.33 \times L \times h^{1.5}$
$80 = 3.33 \times 15h^{1.5}$
$h = 1.4$ ft (w/calculator: 1.6 INV $y^{\times 1.5} = 1.36$ or 1.4)
4 ft height of weir + 1.4 ft head of water = 5.4 ft depth

Triangular weirs, also called V-notch weirs, can have notch angles ranging from 22.5° to 90°, but right-angle notches are the most common [see Figure 8.13(b)].

The formula used to make V-notch (90°) weir calculations is

$$Q = 2.5 \times h^{2.5} \qquad (8.2)$$

where

Q = flow
h = head on weir (measured from bottom of notch to water surface)

EXAMPLE 8.3

Problem: What should be the minimum weir height for measuring a flow of 1200 gpm with a 90° V-notch weir, if the flow is now moving at 4 ft/sec in a 2.5-ft wide rectangular channel?

Solution:

$$\frac{1200 \ gpm}{60 \ sec/min \times 7.48 \ gal/cu \ ft} = 2.67 \ cfs$$

$Q = A \times V$

$2.67 = 2.5 \times d \times 4$

$d = 0.27 \ ft$

$Q = 2.5 \times h^{2.5}$

$2.67 = 2.5 \times h^{2.5}$

$h = 1.03$ (calculator: 1.06 INV $y^{\times 2.5} = 1.027$ or 1.03)

$0.27 \ ft$ (original depth) $+ 1.03$ (head on weir) $= 1.30 \ ft$

It is important to point out that weirs, aside from being operated within their flow limits, must also be operated within the available system head. In addition, the operation of the weir is sensitive to the approach velocity of the water, often necessitating a stilling basin or pond upstream of the weir. Weirs are not suitable for water that carries excessive solid materials or silt, which deposit in the approach channel behind the weir and destroy the conditions required for accurate discharge measurements.

 Important Point: *Accurate flow rate measurements with a weir cannot be expected unless the proper conditions and dimensions are maintained.*

IMPORTANT

8.4.6.2 FLUMES

A *flume* is a specially shaped, constricted section in an open channel (similar to the Venturi tube in a pressure conduit). The special shape of the flume (see Figure 8.14) restricts the channel area and/or changes the channel slope, resulting in an increased velocity and a change in the level of the liquid flowing through the flume. The flume restricts the flow, then expands it in a definite fashion. The flow rate through the flume may be determined by measuring the head on the flume at a single point, usually at some distance downstream from the inlet.

Flumes can be categorized as belonging to one of the three general families, mentioned earlier in Chapter 7, depending upon the state of flow induced—subcritical, critical, or supercritical. Typically, flumes that induce a critical or supercritical state of flow are most commonly used. This is because when critical or supercritical flow occurs in a channel, one head measurement can indicate the discharge rate if it is made far enough upstream so that the flow depth is not affected by the drawdown of the water surface as it achieves or passes through a critical state of flow. For critical or supercritical states of flow, a definitive head-discharge relationship can be established and measured, based on a single head reading. Thus, most commonly encountered flumes are designed to pass the flow from subcritical through critical at or near the point of measurement.

The most common flume used for a permanent wastewater flow-metering installation is called the *Parshall flume,* shown in Figure 8.14.

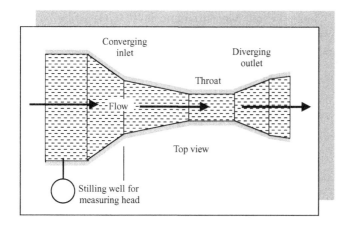

Figure 8.14
Parshall flume.

Formulas for flow through Parshall flumes differ, depending on throat width. The formula below can be used for widths of 1–8 feet and applies to a medium range of flows.

$$Q = 4 \times W \times H_a^{1.52} \times W^{.026} \qquad (8.3)$$

where

Q = flow
H_a = depth in stilling well upstream
W = width of throat

 Note: *Parshall flumes are low-maintenance items.*

IMPORTANT

REFERENCES

Barnes, R. G., Positive Displacement Flowmeters for Liquid Measurement, in *Flow Measurement,* Spitzer, D.W. (ed.). Research Triangle Park, NC: Instrument Society of America, pp. 315–322, 1991.

Brown, A. E., Ultrasonic Flowmeters, in *Flow Measurement,* Spitzer, D. W. (ed.). Research Triangle Park, NC: Instrument Society of America, pp. 415–432, 1991.

Flow Instrumentation: A Practical Workshop on Making Them Work. Sacramento, CA: The Water and Wastewater Instrumentation Testing Association and United States Environmental Protection Agency, Section A, May 16–17, 1991.

Grant, D. M., Open Channel Flow Measurement, in *Flow Measurement,* Spitzer, D. W. (ed.). Research Triangle Park, NC: Instrument Society of America, pp. 252–290, 1991.

Hauser, B. A., *Practical Hydraulics Handbook,* 2nd ed. Boca Raton, FL: Lewis Publishers, p. 91, 1996.

Husain, Z. D., and Sergesketter, M. J., Differential Pressure Flowmeters, in *Flow Measurement,* Spitzer, D. W. (ed.). Research Triangle Park, NC: Instrument Society of America, pp. 119–160, 1991.

Mills, R. C., Magnetic Flowmeters, in *Flow Measurement,* Spitzer, D. W. (ed.). Research Triangle Park, NC: Instrument Society of America, pp. 175–219, 1991.

Nathanson, J. A., *Basic Environmental Technology: Water Supply, Waste Management, and Pollution Control,* 2nd ed. Upper Saddle River, NJ: Prentice Hall, p. 39, 1997.

Oliver, P. D., Turbine Flowmeters, in *Flow Measurement,* Spitzer, D. W. (ed.). Research Triangle Park, NC: Instrument Society of America, pp. 373–414, 1991.

Water Treatment: Principles and Practices of Water Supply Operations, 2nd ed. Denver, CO: American Water Works Association, pp. 449–450, 1995.

Self-Test

8.1 What should be the minimum weir height for measuring a flow of 900 gpm with a 90° V-notch weir, if the flow is now moving at 3 ft/sec in a 2-ft wide rectangular channel?

8.2 A weir 3-ft high extends 10 ft across a rectangular channel in which there are 110 cfs flowing. What is the depth just upstream from the weir?

8.3 A 90° V-notch weir is to be installed in a 30-inch diameter sewer to measure 600 gpm. What head should be expected?

8.4 A secondary clarifier 40-ft wide discharges 1.2 MGD over a weir the width of the tank. How high is the head on the weir (inches)?

8.5 Flow measurement can be based on flow _____, or flow _____.

8.6 List five advantages of differential pressure flowmeters:

8.7 For dirty water operations, an _____ or _____ orifice plate should be used.

8.8 A _____ is a restriction with a relatively long passage with smooth entry and exit.

8.9 _____ have a smooth entry and sharp exit.

8.10 A _____ is a point velocity measuring device.

8.11 _____ send a beam of ultrasonic sound waves through the water to another transducer on the opposite side of the unit.

8.12 Name two devices commonly used to measure flow in open channels.

Hydraulic Machines: Pumps[28]

Few engineered artifacts are as essential as pumps in the development of the culture which our western civilization enjoys. From the smallest to the largest, every facet of our daily lives is served in some measure by such machines. Ancient civilizations, requiring irrigation and essential water supplies, utilized crude forms of pumps which, their design having been refined, are in use even today. Moreover, in today's precise mechanical environment, the types and forms of pump equipment add hundreds of variations to the earlier, simple forms.[29]

TOPICS

Basic Pumping Calculations
Centrifugal Pumps
Pump Control Systems
Protective Instrumentation
Centrifugal Pump Modifications
Positive Displacement Pumps

9.1 INTRODUCTION

Note: *This chapter was prepared using the latest information available; however, it is possible that specific situations may occur*

IMPORTANT *that require actions different from those described in this chapter. In all cases, the recommendations of the equipment manufacturer and its service representatives should be used if contrary to information supplied in this chapter.*

---

[28] Much of the information in this chapter is from Spellman, F. R., Chapter 8 of *The Handbook for Waterworks Operator Certification, Volume 2: Intermediate Level.* Lancaster, PA: Technomic Publishing Co., Inc., 2000. In addition, selected portions contained herein are from notes taken during attendance at the Water Treatment Operators Short Course, presented in August 1999, Virginia Polytechnic Institute and State University (Virginia Tech), Blacksburg, Va.

[29] Garay, P. N., *Pump Application Desk Book.* Lilburn, GA: The Fairmont Press, Inc., p. 1, 1990.

Key Terms Used in This Chapter[30]

ABSOLUTE PRESSURE ▷	The pressure of the atmosphere on a surface. At sea level, a pressure gauge with no external pressure added will read 0 psig. The atmospheric pressure is 14.7 psia. If the gauge reads 15 psig, the absolute pressure will be 15 + 14.7, or 29.7 psia.
ACCELERATION DUE TO GRAVITY (g) ▷	The rate at which a falling body gains speed. The acceleration due to gravity is 32 feet/second/second. This simply means that a falling body or fluid will increase the speed at which it is falling by 32 feet/second every second that it continues to fall.
ATMOSPHERIC PRESSURE ▷	The pressure exerted on a surface area by the weight of the atmosphere is atmospheric pressure, which at sea level is 14.7 psi, or one atmosphere. At higher altitudes, the atmospheric pressure decreases. At locations below sea level, the atmospheric pressure rises (see Table 9.1).
CAVITATION ▷	An implosion of vapor bubbles in a liquid inside a pump caused by a rapid local pressure decrease occurring mostly close to or touching the pump casing or impeller. As the pressure reduction continues, these bubbles collapse or implode.

[30] Definitions are from Spellman, F. R., *The Science of Water.* Lancaster, PA: Technomic Publishing Co., Inc., 1998; Wahren, U., *Practical Introduction to Pumping Technology.* Houston: Gulf Publishing Company, 1997; Hauser, B. A., *Hydraulics for Operators.* Boca Raton, FL: Lewis Publishers, 1993; and *Basic Science Concepts and Applications: Principles and Practices of Water Supply Operations,* 2nd ed., Denver: American Water Works Association, 1995.

CAVITATION (continued)	Cavitation may produce noises that sound like pebbles rattling inside the pump casing and may also cause the pump to vibrate and to lose hydrodynamic efficiency. This effect contrasts boiling, which happens when heat builds up inside the pump. Continued serious cavitation may destroy even the hardest surfaces. Avoiding cavitation is one of the most important pump design criteria. Cavitation limits the upper and lower pump sizes, as well as the pump's peripheral impeller speed.
CRITICAL SPEED	At this speed, a pump may vibrate enough to cause damage. Pump manufacturers try to design pumps with the first critical speed at least 20 percent higher or lower than rated speed. Second and third critical speeds usually don't apply in pump usage.
CROSS-SECTIONAL AREA (A)	The area perpendicular to the flow that the liquid in a channel or pipe occupies (see Figure 9.1).
DISPLACEMENT	The capacity, or flow, of a pump is its displacement. This measurement,

TABLE 9.1. Atmospheric Pressure versus Altitude.

Altitude	Barometric Pressure	Equivalent Head	Maximum Practical Suction Lift (Water)
−1000 ft	15.2 psi	35.2 ft	22 ft
Sea Level	14.7 psi	34.0 ft	21 ft
1500 ft	13.9 psi	32.2 ft	20 ft
3000 ft	13.2 psi	30.5 ft	18 ft
5000 ft	12.2 psi	28.3 ft	16 ft
7000 ft	11.3 psi	26.2 ft	15 ft

Note: Water Temperature = 75°F.

DISPLACEMENT (*continued*)	primarily used with positive-displacement pumps, is measured in units such as gallons, cubic inches, and liters.
DENSITY	Recall that this term refers to the mass per unit volume measured in pounds per cubic foot at 68°F or in grams per milliliter at 4°C.
DISCHARGE PRESSURE	The pressure measured at the pump's discharge nozzle. Measurements may be stated in psig, kg/cm^2, bars, or kilopascals.
ENERGY	The ability to do work.
Potential Energy	Energy due to the liquid's location or condition.
Kinetic Energy	Energy of motion.
FLOW	The volume, quantity, or amount of fluid that passes a point in a given amount of time—flow can be viewed as a moving volume. It is measured in million gallons/day, gallons/day, and cubic feet/second. In most hydraulics calculations, the flow is expressed in cubic feet/second, cfs. To obtain cubic feet per second when flow is given in million gallons per day, multiply by 1.55 cfs/MGD.

$$Q, \text{cfs} = \text{MGD} \times 1.55 \text{ cfs/MGD}$$
$$(9.1)$$

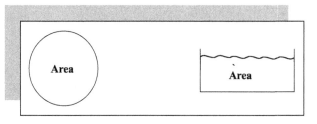

Figure 9.1
Cross-sectional area.

FLOW
(*continued*)

In pumping hydraulics, flow refers to the liquid that enters the pump's suction nozzle. Flow (*Q*) measurements are U.S. gallons per minute (Usgpm or gpm) and can be converted as follows:

- imperial gallons per minute = Usgpm × 1.200

- cubic meters per hour (m^3/hr) = Usgpm × 0.227

- liters per second (L/sec) = Usgpm × 0.063

- barrels per day (1 barrel = 42 gal) = Usgpm × 34.290

IMPORTANT

Important Point: *The pump's flow capacity varies with impeller width, impeller diameter, and pump revolutions per minute (rpm).*

HEAD

The energy a liquid possesses at a given point or a pump must supply to move a liquid to a given location. Head is expressed in feet.

Cut-Off Head

The head at which the energy supplied by a pump and the energy required to move the liquid to a specified point are equal and no discharge at the desired point will occur.

Discharge Head

Measured in feet or meters, the discharge head is the same as the discharge pressure converted into the height of a liquid column.

Friction Head

The amount of energy in feet that is necessary to overcome the resistance to flow that occurs in the pipes and fixtures through which the liquid is flowing through.

Pressure Head	The vertical distance a given pressure can raise a liquid. For example, if a liquid has a pressure of 1 pound per square inch (psi), the liquid will rise to a height of 2.31 feet.
Pump Head	The energy in feet that a pump supplies to the fluid.
Static Head	The energy in feet required to move a fluid from the supply tank to the discharge point (see Figure 9.2).

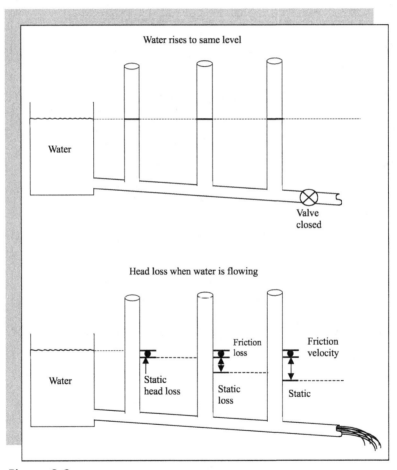

Figure 9.2
Head loss in non-pumping system.

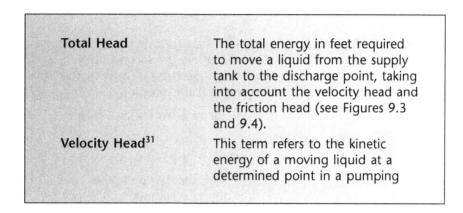

| Total Head | The total energy in feet required to move a liquid from the supply tank to the discharge point, taking into account the velocity head and the friction head (see Figures 9.3 and 9.4). |
| Velocity Head[31] | This term refers to the kinetic energy of a moving liquid at a determined point in a pumping |

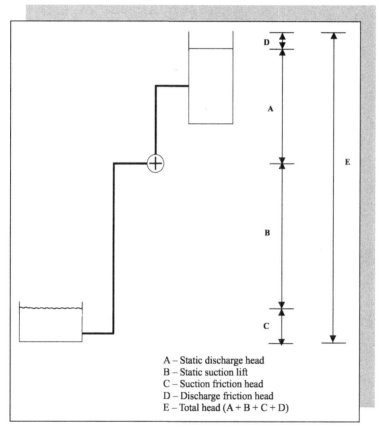

A – Static discharge head
B – Static suction lift
C – Suction friction head
D – Discharge friction head
E – Total head (A + B + C + D)

Figure 9.3
Head components for suction lift system.

[31] Wahren, U., *Practical Introduction to Pumping Technology.* Houston: Gulf Publishing Company, pp. 6–7, 1997.

station. The expression for velocity head is in feet per second (ft/sec) or meters per second (m/sec). The mathematical expression is

$$\text{Velocity head } (h_v) = V^2/2g$$

(9.2)

where:

V = liquid velocity in a pipe
g = gravity acceleration, influenced by both altitude and latitude. At sea level and 45° latitude, it is 32.17 ft/sec/sec.

Note: *If the pump inlet nozzle and discharge nozzle are of equal size, then this term is normally zero.*

IMPORTANT

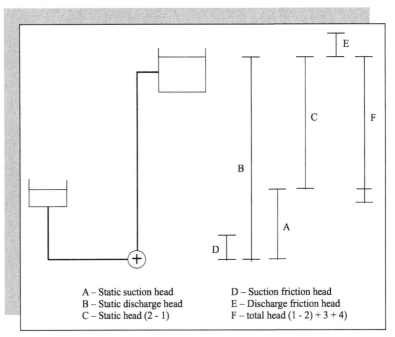

A – Static suction head
B – Static discharge head
C – Static head (2 - 1)

D – Suction friction head
E – Discharge friction head
F – total head (1 - 2) + 3 + 4)

Figure 9.4
Head components for suction head type system.

Suction Head	The total head in feet on the suction or supply side of the pump when the supply is located above the center of the pump.
Discharge Head	The total head in feet on the discharge side of the pump.
Suction Lift	The total head in feet on the suction or supply side of the pump, when the supply is located below the center of the pump.
GAUGE PRESSURE	As the name implies, pressure gauges show gauge pressure (psig), which is the pressure exerted on a surface minus the atmospheric pressure. Thus, if the absolute pressure in a pressure vessel is 150 psia, the pressure gauge will read $150 - 14.7$, or 135.3 psig
HORSEPOWER	The work a pump performs while moving a determined amount of liquid at a given pressure is horsepower (hp).
Hydraulic Horsepower	(whp) measures pump output. This term also is known as **water horsepower.**
Brake Horsepower	(bhp) measures the input horsepower delivered to the pump shaft.
MINIMUM FLOW BYPASS	Refers to a pipe that leads from the pump discharge piping back into the pump suction system. A pressure control, or flow control, valve opens this line when the pump discharge flow approaches the pump's minimum flow value. The purpose is to protect the pump from damage.
MINIMUM FLOW	The lowest continuous flow at which a manufacturer will guarantee a pump's performance is the pump's minimum flow.

POWER	Use of energy to perform a given amount of work in a specified length of time. In most cases, this is expressed in terms of horsepower.
SPECIFIC GRAVITY	Recall that if we divide the weight of a body by the weight of an equal volume of water at 68°F, we get specific gravity (sp gr). If the data are in grams per milliliter, the specific gravity of a body of water is the same as its density at 4°C.
VACUUM	Any pressure below atmospheric pressure is a partial vacuum. The expression for vacuum is in inches or millimeters of mercury (Hg). Full vacuum is at 30 in. Hg. To convert inches to millimeters, multiply inches by 25.4.
VAPOR PRESSURE	At a specific temperature and pressure, a liquid will boil. The point at which the liquid begins to boil is the liquid's vapor pressure point. The vapor pressure (*vp*) will vary with changes in either temperature or pressure, or both.
VELOCITY (*V*)	The speed of the fluid moving through a pipe or channel. It is normally expressed in feet per second (fps).
VOLUMETRIC EFFICIENCY	Obtained by dividing a pump's actual capacity by the calculated displacement to get volumetric efficiency. The expression is primarily used in connection with positive displacement pumps.
WORK	Using energy to move an object a distance. It is usually expressed in foot-pounds.

In the chapter opening, Garay points out that "few engineered arti-facts are as essential as pumps in the development of the culture which our western civilization enjoys." This statement is germane to any dis-cussion about pumps simply because man has always needed to move water from one place to another against the forces of nature.

In his earliest efforts, man used one of the primary forces of nature, gravity, to assist him in moving water from one place to another. Gravity only works, of course, if the water is moved downhill on a grade. Man soon discovered that if he built up an accumulation of water behind the water source (e.g., behind a dam), pressure moved the water further. But when pressure is dissipated by various losses (e.g., friction loss), when wa-ter in low-lying areas is needed in higher areas, the energy needed to move that water must be created. Simply, some type of pump is needed.

In 287 BC, Archimedes (Greek mathematician and physicist) invented the screw pump (see Figure 9.5).

The Roman emperor, Nero, around 100 AD, is often given credit for developing the piston pump. In operation, the piston pump displaces

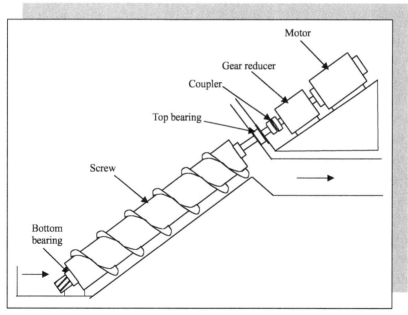

Figure 9.5
Archimedes' screw pump.

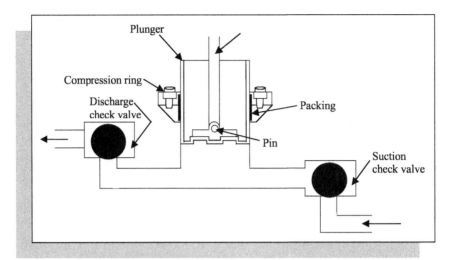

Figure 9.6
Piston pump.

volume after volume of water with each stroke (see Figure 9.6). The piston pump has two basic problems: (1) its size limits capacity and (2) it is a high energy consumer.

It was not until the nineteenth century that pumping technology took a leap forward from its rudimentary beginnings. The first fully functional centrifugal pumps were developed in the 1800s. Centrifugal pumps can move great quantities of water with much smaller units than the pumps previously in use.

The pump is a type of hydraulic machine. Pumps convert mechanical energy into fluid energy. Whether it is taken from groundwater or a surface water body, from one unit treatment process to another, or to the storage tank for eventual final delivery through various sizes and types of pipes to the customer, **pumps** are the usual source of energy necessary for the conveyance of water. The only exception may be, of course, where the source of energy is supplied entirely by gravity. Waterworks/wastewater maintenance operators must, therefore, be familiar with pumps, pump characteristics, pump operation, and pump maintenance.

"There are three general requirements of pump and motor combinations. These requirements are (1) reliability (2) adequacy and (3) economy. **Reliability** is generally obtained by installing in duplicate the very best equipment available and by the use of an auxiliary power source. **Adequacy** is obtained by securing liberal sizes of pumping equipment. **Economics** can be achieved by taking into account the life and depreciation, first cost, standby charges, interest and operating costs."[32]

During the past several years, it has become more evident that many waterworks and wastewater facilities have been unable to meet their optimum supply and/or treatment requirements for one of three reasons:

1. Untrained operations and maintenance staff

2. Poor plant maintenance

3. Improper plant design

This chapter addresses the basics, the fundamentals of pumping technology, and does not address the three important items listed above. However, the three items listed above are discussed in greater detail in the *Pumping* volume of this series. The intent of this volume is only to provide the water/wastewater maintenance operator with a basic knowledge of the technology and principles of operation of the centrifugal and positive-displacement pumps. (Note that this volume sets the foundation for the material presented in *Pumping.*)

It is unlikely that the material in this chapter is sufficient to allow an inexperienced person to perform all the maintenance requirements of centrifugal/positive displacement pumping facilities. However, the information contained in this chapter, when combined with in-plant experience and the principles presented in *Pumping,* should achieve the desired result; that is, better performance through maintenance of plant flexibility and reductions in unscheduled shutdowns of critical pumping facilities. More important, and more immediate to the maintenance operator-in-training, is the certainty that the material contained within this chapter should better enable the operator to correctly answer certification examination questions dealing with basic hydraulics, pump basics, and centrifugal and/or positive-displacement pumps in general.

---

[32]"Texas Manual." *Manual of Water Utility Operations,* 8th ed. Austin TX: Texas Utilities Association, p. 372, 1988 (emphasis added).

9.2 BASIC PUMPING CALCULATIONS

Basic calculations are a fact of life that the water/wastewater maintenance operator soon learns, and hopefully learns well enough to use as required to operate a water/wastewater facility correctly.

In the following sections, the basic calculations used frequently in water hydraulic and pumping applications are discussed. Only those types of "basic" calculations that the water/wastewater maintenance operator may be required to know for operational and certification purposes are included. Calculations for pump-specific speed, suction-specific speed, affinity formulae, and other advanced calculations are, although at a higher technical level, also covered in this section.

9.2.1 VELOCITY OF A FLUID THROUGH A PIPELINE

The speed or velocity of a fluid flowing through a channel or pipeline is related to the cross-sectional area of the pipeline and the quantity of water moving through the line.

For example, if the diameter of a pipeline is reduced, then the velocity of the water in the line must increase to allow the same amount of water to pass through the line.

$$\text{Velocity } (V) \text{ fps} = \frac{\text{Flow } (Q) \text{ cfs}}{\text{Cross-Sectional Area } (A) \text{ ft}^2} \qquad (9.3)$$

EXAMPLE 9.1

Problem: If the flow through a 2-ft diameter pipe is 9 MGD, the velocity is

---

Solution:

$$\text{Velocity, fps} = \frac{9\text{ MGD} \times 1.55\text{ cfs/MGD}}{0.785 \times 2\text{ ft} \times 2\text{ ft}}$$

$$V = \frac{14\text{ cfs}}{3.14\text{ ft}^2}$$

$$V = 4.5\text{ fps (rounded)}$$

EXAMPLE 9.2

Problem: If the same 9-MGD flow used in Example 9.1 is transferred to a pipe with a 1-foot diameter, the velocity would be

Solution:

$$\text{Velocity, fps} = \frac{9\text{ MGD} \times 1.55\text{ cfs/MGD}}{0.785 \times 1\text{ ft} \times 1\text{ ft}}$$

$$V = \frac{14\text{ cfs}}{0.785\text{ ft}^2}$$

$$V = 17.8\text{ fps (rounded)}$$

These sample problems show that if the cross-sectional area is decreased, the velocity of the flow must be increased. Mathematically, we can say that the velocity and cross-sectional area are inversely proportional when the amount of flow (Q) is constant (see Figure 9.7)

$$\textbf{Area}_1 \times \textbf{Velocity}_1 = \textbf{Area}_2 \times \textbf{Velocity}_2 \qquad (9.4)$$

Note: *The concept just explained is extremely important in the operation of a centrifugal pump and will be discussed later.*

IMPORTANT

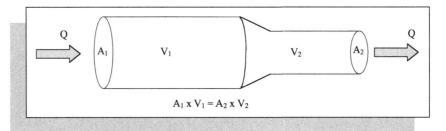

Figure 9.7
Area-velocity relationship.

9.2.2 PRESSURE-VELOCITY RELATIONSHIP

A relationship similar to that of velocity and cross-sectional area exists for velocity and pressure. As the velocity of flow in a full pipe increases, the pressure of the liquid decreases (see Figure 9.8). This relationship is

$$P_1 \times V_1 = P_2 \times V_2 \tag{9.5}$$

EXAMPLE 9.3

Problem: If the flow in a pipe has a velocity of 3 fps and a pressure of 4 psi and the velocity of the flow increases to 4 fps, the pressure will be

Solution:

$$P_1 \times V_1 = P_2 \times V_2$$
$$4 \text{ psi} \times 3 \text{ fps} = P_2 \times 4 \text{ fps}$$

Rearranging

$$P_2 = \frac{4 \text{ psi} \times 3 \text{ fps}}{4 \text{ fps}}$$
$$P_2 = \frac{12 \text{ psi}}{4}$$
$$P_2 = 3 \text{ psi}$$

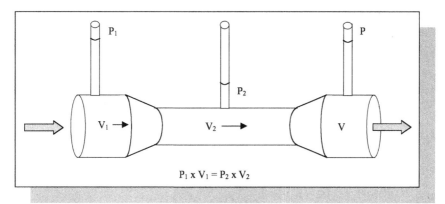

Figure 9.8
Pressure-velocity relationship.

Again, this is another important hydraulics principle that is very important to the operation of a centrifugal pump.

9.2.3 STATIC HEAD

Pressure at a given point originates from the height, or depth, of water above it. It is this pressure, or *head,* that gives the water energy and causes it to flow.

By definition, *static head* is the vertical distance the liquid travels from the supply tank to the discharge point. This relationship is shown as

Static Head, ft = Discharge Level, ft − Supply Level, ft

(9.6)

In many cases, it is desirable to separate the static head into two separate parts: (1) the portion that occurs before the pump (suction head or suction lift) and (2) the portion that occurs after the pump (discharge head). When this is done, the center (or datum) of the pump becomes the reference point.

9.2.4 STATIC SUCTION HEAD

Static suction head refers to when the supply is located above the pump datum.

Static Suction Head, ft = Supply Level, ft = Pump Level, ft

$$(9.7)$$

9.2.5 STATIC SUCTION LIFT

Static suction lift refers to when the supply is located below the pump datum.

Static Suction Lift, ft = Pump Level, ft − Supply Level, ft

$$(9.8)$$

9.2.6 STATIC DISCHARGE HEAD

Static Discharge Head, ft = Discharge Level, ft
− Pump Datum, ft $$(9.9)$$

If the total static head is to be determined after calculating the static suction head or lift and static discharge head, individually, there are two separate calculations, depending on whether there is a suction head or a suction lift.

For Suction Head:

Total Static Head = Static Discharge Head, ft
− Static Suction Lift, ft $$(9.10)$$

For Suction Lift:

Total Static Head, ft = Static Discharge Head, ft
+ Static Suction Lift, ft $$(9.11)$$

EXAMPLE 9.4

Problem: Refer to Figure 9.9.

Solution:

Step 1: Static Suction Lift, ft = Pump Level, ft − Supply Level, ft

Static Suction Lift, ft = 128 ft − 121 ft

= 7 ft

Step 2: Static Discharge Head, ft = Discharge Level, ft + Static Suction Lift, ft

= 145 ft − 128 ft

= 17 ft

Step 3: Total Static Head, ft = Static Discharge Head, ft + Static Suction Lift

= 17 ft + 7 ft

= 24 ft

or

Total Static Head, ft = Discharge Level, ft − Supply Level, ft

= 145 ft − 121 ft

= 24 ft

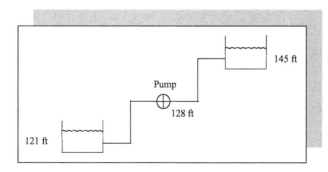

Figure 9.9
For Example 9.4.

EXAMPLE 9.5

Problem: Refer to Figure 9.10.

Solution:

Step 1: Static Suction Head, ft = Supply Level, ft − Pump Level, ft

= 124 ft − 117 ft

= 7 ft

Step 2: Static Discharge Head, ft = Discharge Level, ft − Pump Level, ft

= 141 ft − 117 ft

= 24 ft

Step 3: Total Static Head, ft = Static Discharge Head, ft − Static Suction Head

= 24 ft − 7 ft

= 17 ft

or

Total Static Head, ft = Discharge Level, ft − Supply Level, ft

= 141 ft − 124 ft

= 17 ft

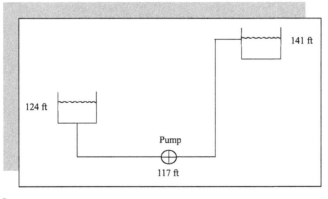

Figure 9.10
For Example 9.5.

9.2.7 FRICTION HEAD

Various formulae calculate friction losses. Hazen-Williams wrote one of the most common for smooth steel pipe. Usually, we do not need to calculate the friction losses, because handbooks such as the *Hydraulic Institute Pipe Friction Manual* tabulated these long ago. This important manual also shows velocities in different pipe diameters at varying flows, as well as the resistance coefficient (K) for valves and fittings.[33]

Friction head (in feet) is the amount of energy used to overcome resistance to the flow of liquids through the system. It is affected by the length and diameter of the pipe, the roughness of the pipe, and the velocity head. It is also affected by the physical construction of the piping system. The number of and types of elbows, valves, Ts, etc., will greatly influence the friction head for the system. These must be converted to their equivalent length of pipe and included in the calculation.

$$\text{Friction Head, ft} = \text{Roughness Factor}, f \times \frac{\text{Length}}{\text{Diameter}} \times \frac{\text{Velocity}^2}{2g}$$

$$(9.12)$$

The *roughness factor (f)* varies with length and diameter as well as the condition of the pipe and the material from which it is constructed; it is normally in the range of 0.01–0.04.

 IMPORTANT ***Important Point:*** *For centrifugal pumps, good engineering practice is to try to keep velocities in the suction pipe to 3 ft/sec or less. Discharge velocities higher than 11 ft/sec may cause turbulent flow and/or erosion in the pump casing.*

EXAMPLE 9.6

Problem: What is the friction head in a system that uses 150 ft of 6-inch diameter pipe, when the velocity is 3 fps? The system's valving is equivalent

---

[33] Wahren, U. *Practical Introduction to Pumping Technology.* Houston: Gulf Publishing Company, p. 9, 1997.

to an additional 75 feet of pipe. Reference material indicates a roughness factor (*f*) of 0.025 for this particular pipe and flow rate.

Solution:

$$\text{Friction Head, ft} = \text{Roughness factor}$$
$$\times \frac{\text{Length}}{\text{Diameter}} \times \frac{\text{Velocity}}{2g}$$

$$\text{Friction Head, ft} = 0.025$$
$$\times \frac{(150 \text{ ft} + 75 \text{ ft})}{0.5 \text{ ft}}$$
$$\times \frac{(3 \text{ fps})^2}{2 \times 32 \text{ ft/sec}^2}$$

$$\text{Friction Head, ft} = 0.025 \times \frac{225 \text{ ft}}{0.5 \text{ ft}}$$
$$\times \frac{9 \text{ ft}^2/\text{s}^2}{64 \text{ ft/sec}^2}$$

$$\text{Friction Head, ft} = 0.025 \times 450$$
$$\times 0.140 \text{ ft}$$

$$\text{Friction Head, ft} = 1.58 \text{ ft}$$

It is also possible to compute friction head using tables. Friction head can also be determined on both the suction side of the pump and the discharge side of the pump. In each case, it is necessary to determine

1. The length of pipe

2. The diameter of the pipe

3. Velocity

4. Pipe equivalent of valves, elbows, Ts, etc.

9.2.8 VELOCITY HEAD

Velocity head is the amount of head or energy required to maintain a stated velocity in the suction and discharge lines. The design of most pumps makes the total velocity head for the pumping system zero.

Note: *Velocity head only changes from one point to another on a pipeline if the diameter of the pipe changes.*

IMPORTANT

Velocity head and total velocity head are determined by

$$\text{Velocity Head, ft} = \frac{(\text{Velocity})^2}{2g} \qquad (9.13)$$

$$\text{Total Velocity Head, ft} = \text{Velocity Head Discharge, ft} - \text{Velocity Head Suction, ft} \qquad (9.14)$$

EXAMPLE 9.7

Problem: What is the velocity head for a system that has a velocity of 5 fps?

Solution:

$$\text{Velocity Head, ft} = \frac{(\text{Velocity})^2}{2 \times \text{Acceleration due to gravity}}$$

$$\text{Velocity Head, ft} = \frac{(5 \text{ fps})^2}{2 \times 32 \text{ ft/sec}^2}$$

$$\text{Velocity Head, ft} = \frac{(25 \text{ ft}^2/\text{sec}^2}{64 \text{ ft/sec}^2}$$

$$\text{Velocity Head, ft} = 0.39 \text{ ft}$$

Note: *There is no velocity head in a static system. The water is not moving.*

IMPORTANT

9.2.9 TOTAL HEAD

Total head is the sum of the static, friction, and velocity head.

Total Head, ft = Static Head, ft + Friction Head, ft
+ Velocity Head, ft *(9.15)*

9.2.10 CONVERSION OF PRESSURE HEAD

Pressure is directly related to the head. If liquid in a container subjected to a given pressure is released into a vertical tube, the water will rise 2.31 ft for every pound per square inch of pressure.

To convert pressure to head in feet:

Head, ft = Pressure, psi × 2.31 ft/psi *(9.16)*

This calculation can be very useful in cases where liquid is moved through another line that is under pressure. Because the liquid must overcome the pressure in the line it is entering, the pump must supply this additional head.

EXAMPLE 9.8

Problem: A pump is discharging to a pipe that is full of liquid under a pressure of 20 psi. The pump and piping system has a total head of 97 ft. How much additional head must the pump supply to overcome the line pressure?

Solution:

$$Head, ft = Pressure, psi \times 2.31 \ ft/psi$$
$$= 20 \ psi \times 2.31 \ ft/psi$$
$$= 46 \ ft \ (rounded)$$

Note: *The pump must supply an additional head of 46 ft to overcome the internal pressure of the line.*

IMPORTANT

9.2.11 HORSEPOWER

The unit of work is **foot pound;** the amount of **work** required to lift a one-pound object one foot off the ground (ft-lb). For practical purposes, the amount of work being done is considered. It is more valuable, obviously, to be able to work faster; that is, for economic reasons, we consider the rate at which work is being done (i.e., power or foot pound/second). At some point, the horse was determined to be the ideal work animal; it could move 550 pounds one foot, in one second, considered to be equivalent to one horsepower.

550 ft lb/sec = 1 Horsepower (hp)

or

33,000 ft lb/min = 1 Horsepower (hp)

A pump performs work while it pushes a certain amount of water at a given pressure. The two basic terms for horsepower are (1) hydraulic horsepower and (2) brake horsepower.

9.2.11.1 HYDRAULIC (WATER) HORSEPOWER (whp)

A pump has power because it does work. A pump lifts water (which has weight) a given distance in a specific amount of time (ft/lb/min).

One hydraulic (water) horsepower (whp) provides the necessary power to lift the water to the required height; it equals the following:

- 550 ft-lb/sec

- 33,000 ft-lb/min

- 2545 British thermal units per hour (Btu/hr)

💧 0.746 kw

💧 1.014 metric hp

To calculate the hydraulic horsepower (whp) using flow in gpm and head in feet, use the following formula for centrifugal pumps:

$$\text{whp} = \frac{\textbf{flow (in gpm)} \times \textbf{head (in ft)} \times \textbf{specific gravity}}{\textbf{3960}} \qquad (9.17)$$

Note that 3960 is derived by dividing 33,000 ft-lb by 8.34 lb/gal = 3960.

9.2.11.2 BRAKE HORSEPOWER (BHP)

A water pump does not operate alone. It is driven by the motor, and electrical energy drives the motor. Brake horsepower is the horsepower applied to the pump.

A pump's brake horsepower (BHP) equals its hydraulic horsepower divided by the pump's efficiency. (Note: Neither the pump nor its prime mover (motor) is 100% efficient. There are friction losses within both of these units, and it will take **more** horsepower applied to the pump to get the required amount of horsepower to move the water, and even more horsepower applied to the motor to get the job done.[34]) Thus, the BHP formulas become

$$\text{BHP} = \frac{\textbf{flow (in gpm)} \times \textbf{head (in ft)} \times \textbf{specific gravity}}{\textbf{3960} \times \textbf{efficiency}} \qquad (9.18)$$

IMPORTANT

Important Points: *(1)* **Water Horsepower** *(whp) is the power necessary to lift the water to the required height;* **Brake Horsepower** *(BHP) is the horsepower applied to the pump; (3)* **Motor Horsepower** *(hp) is the horsepower applied to the motor; and (4)* **Efficiency** *is the power produced by the unit, divided by the power used in operating the unit.*

---

[34] Hauser, B. A., *Hydraulics for Operators.* Boca Raton, FL: Lewis Publishers, p. 38, 1993.

9.2.12 SPECIFIC SPEED[35]

The capacity of flow rate of a centrifugal pump is governed by the impeller thickness. For a given impeller diameter, the deeper the vanes, the greater the capacity of the pump.

For desired flow rate or a desired discharge head, there will be one optimum impeller design. The impeller that is best for developing a high discharge pressure will have different proportions from an impeller designed to produce a high flow rate. The quantitative index of this optimization is called *specific speed (n_s)*. The higher the specific speed of a pump, the higher its efficiency.

An impeller's specific speed is its speed when pumping 1 gpm of water at a differential head of 1 ft. The following formula is used to determine specific speed (where H is at the best efficiency point):

$$N_s = \frac{\text{rpm} \times Q^{0.5}}{H^{0.75}} \qquad (9.19)$$

where

    rpm = revolutions per minute
    Q = flow (in gpm)
    H = head (in ft)

Pump-specific speeds vary between pumps. Although no absolute rule sets the specific speed for different kinds of centrifugal pumps, the following rule of thumb for Ns can be used:

● volute, diffuser, and vertical turbine = 500–5000

● mixed flow = 5000–10,000

● propeller pumps = 9000–15,000

---

[35] From Lindeburg, M. R., *Civil Engineering Reference Manual,* 4th ed. San Carlos, CA: Professional Publications, Inc., pp. 4–10, 1986 and Wahren, U., *Practical Introduction to Pumping Technology.* Houston, TX: Gulf Publishing Co., pp. 16–18, 1997.

9.2.13 SUCTION-SPECIFIC SPEED[36]

Suction-specific speed (n_{ss}), also an impeller design characteristic, is an index of the suction characteristics of the impeller (i.e., the suction capacities of the pump). For practical purposes, n_{ss} ranges from about 3000 to 15,000. The limit for the use of suction-specific speed impellers in water is approximately 11,000. The following equation expresses n_{ss}:

$$n_{ss} = \frac{\text{rpm} \times Q^{0.5}}{\text{NPSHR}^{0.75}} \qquad (9.20)$$

where

$$\text{rpm} = \text{revolutions per minute}$$
$$Q = \text{flow in gpm}$$
$$\text{NPSHR} = \text{net positive suction head required}$$

Ideally, n_{ss} should be approximately 7900 for single suction pumps and 11,200 for double suction pumps. (The value of Q in equation 9.20 should be halved for double suction pumps.)

9.2.14 AFFINITY LAWS—CENTRIFUGAL PUMPS

Most parameters (impeller diameter, speed, and flow rate) determining a pump's performance can vary. If the impeller diameter is held constant and the speed varied, the following ratios are maintained with no change of efficiency (Note: because of inexact results, some deviations may occur in the calculations.):

$$Q_2/Q_1 = D_2/D_1 \qquad (9.21)$$
$$H_2/H_1 = (D_2/D_1)^2 \qquad (9.22)$$
$$\text{bhp}_2/\text{bhp}_1 = (D_2/D_1)^3 \qquad (9.23)$$

---

[36] From Wahren, U., *Practical Introduction to Pumping Technology.* Houston, TX: Gulf Publishing Company, p. 17, 1997.

where:

Q = flow
H_1 = head before change
H_2 = head after change
bhp = brake horsepower
D_1 = impeller diameter before change
D_2 = impeller diameter after change

The relation between speed (N) changes are as follows:

$$Q_2/Q_1 = N_2/N_1 \qquad\qquad (9.24)$$
$$H_2/H_1 = (N_2/N_1)^2 \qquad\qquad (9.25)$$
$$\mathbf{bhp_2/bhp_1 = (N_2/N_1)^3} \qquad\qquad (9.26)$$

where:

N_1 = initial rpm
N_2 = changed rpm

EXAMPLE 9.9

Problem: Change an 8-in. diameter impeller for a 9-in. diameter impeller, and find the new flow (Q), head (H), and brake horsepower (bhp) where the 8-in. diameter data are

$$Q_1 = 340 \text{ rpm}$$
$$H_1 = 110 \text{ ft}$$
$$bhp_1 = 10$$

Solution: The 9-in. impeller diameter data will be as follows:

$$Q_2 = 340 \times 9/8 = 383 \text{ gpm}$$
$$H_2 = 110 \times (9/8)^2 = 139 \text{ ft}$$
$$bhp_2 = 10 \times (9/8)^3 = 14$$

9.2.15 NET POSITIVE SUCTION HEAD (NPSH)[37]

NPSHR, as stated earlier, is the net positive suction head required. NPSHR and NPSH (net positive suction head) are terms that are important in pumping technology. *Net positive suction head* (NPSH) is different from both suction head and suction pressure, and this important point tends to be confusing to those first introduced to the term. For instance, when an impeller in a centrifugal pump spins, the motion creates a partial vacuum in the impeller eye. The net positive suction head available is the height of the column of liquid that will fill this partial vacuum without allowing the liquid's vapor pressure to drop below its flash point. That is, this is the NPSH required (NPSHR) for the pump to function properly.

The Hydraulic Institute[38] defines NPSH as "the total suction head in feet of liquid absolute determined at the suction nozzle and referred to datum less the vapor pressure of the liquid in feet absolute." This defines the NPSH available (NPSHA) for the pump. (Note: NPSHA is the actual water energy at the inlet.) The important point: a pump will run satisfactorily if the NPSHA equals or exceeds the NPSHR. Most authorities recommend the NPSHA be at least 2 ft absolute or 10 percent larger than the NPSHR, whichever number is larger.

IMPORTANT

Note: *NPSHR, simply explained: Contrary to popular belief, water is not sucked into a pump. A positive head (normally atmospheric pressure) must push the water into the impeller (i.e., flood the impeller). NPSHR is the minimum water energy required at the inlet by the pump for satisfactory operation. The pump manufacturer usually specifies NPSHR.*

It is important to point out that if NPSHA is less than NPSHR, the water will cavitate. *Cavitation* is the vaporization of fluid within the

[37] From Lindeburg, M. R., *Civil Engineering Reference Manual*, 4th ed. San Carlos, CA: Professional Publications, Inc., pp. 4–5, 1986, and Wahren, U., *Practical Introduction to Pumping Technology*. Houston, TX: Gulf Publishing Company, pp. 66–68, 1997.

[38] *Hydraulic Institute Complete Pump Standards*, 4th ed. Cleveland: Hydraulic Institute, p. 11, 1994, and *The Hydraulic Institute Engineering Data Book*, 2nd ed. Cleveland: Hydraulic Institute, 1990.

casing or suction line. If the water pressure is less than the vapor pressure, pockets of vapor will form. As vapor pockets reach the surface of the impeller, the local high water pressure will collapse them, causing noise, vibration, and possible structural damage to the pump.

9.2.15.1 CALCULATING NPSHA

The following two example demonstrate how to calculate NPSH for two real world situations: (1) determining NPSHA for an open-top water tank or a municipal water storage tank with a roof and correctly sized vent; and (2) the NPSHA for a suction lift from an open reservoir.

9.2.15.2 NPSHA: ATMOSPHERIC TANK

The following calculation may be used for an open-top water tank or a municipal water storage tank with a roof and correctly sized vent as shown in Figures 9.11 and 9.12.

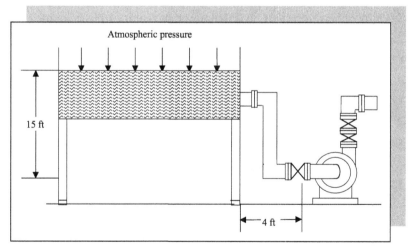

Figure 9.11
Open atmospheric tank.

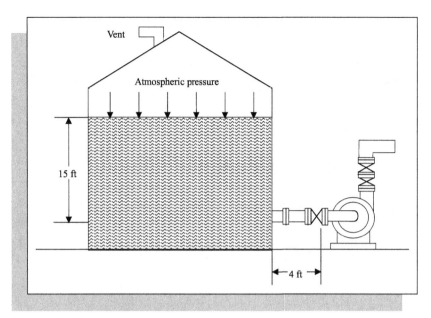

Figure 9.12
Roofed water storage tank.

The formula for calculating NPSHA is

$$\text{NPSHA} = P_a + h - P_v - h_e - h_f \qquad (9.27)$$

where:

P_v = vapor pressure in absolute of water at given temperature
P_a = atmospheric pressure in absolute or pressure of gases against surface of water
h = weight of liquid column from surface of water to center of pump suction nozzle in feet absolute
h_e = entrance losses in feet absolute
h_f = friction losses in suction line in feet absolute

EXAMPLE 9.10

Problem: Given the following, find the NPSHA.

liquid = water
$t = 60°F$

$$sp\ gr = 1.0$$
$$P_v = 0.256\ psia\ (0.6\ ft)$$
$$h_e = 0.4\ ft$$
$$h = 15\ ft$$
$$h_f = 2\ ft$$
$$P_a = 14.7\ psia\ (34\ ft)$$

Solution:

$$NPSHA = 34\ ft + 15\ ft - 0.6\ ft$$
$$- 0.4\ ft - 2\ ft$$
$$NPSHA = 46\ ft$$

9.2.15.2.2 NPSHA: SUCTION LIFT FROM OPEN RESERVOIR (FIGURE 9.13)

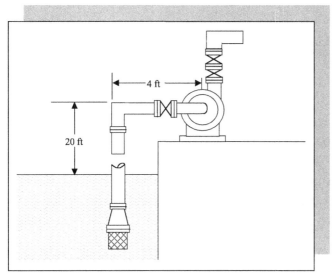

Figure 9.13
Suction lift from open reservoir.

EXAMPLE 9.11

Problem: Find the NPSHA, where:

$$liquid = water$$
$$t = 60°F$$
$$sp\ gr = 1.0$$
$$P_v = 0.256\ psia\ (0.6\ ft)$$
$$Q = 120\ gpm$$
$$h_e = 0.4\ ft$$
$$h_f = 2\ ft$$
$$h = -20\ ft$$
$$P_a = 14.7\ psia\ (34\ ft)$$

Solution:

$$NPSHA = 34\ ft - 20\ ft - 0.6\ ft$$
$$- 0.4\ ft - 2\ ft$$
$$NPSHA = 11\ ft$$

9.2.16 PUMPS IN SERIES AND PARALLEL

Parallel operation is obtained by having two pumps discharging into a common header. This type of connection is advantageous when the system demand varies greatly. An advantage of operating pumps in parallel is that when two pumps are online, one can be shut down during low demand. This allows the remaining pump to operate close to its optimum efficiency.

Series operation is achieved by having one pump discharge into the suction of the next. This arrangement is used primarily to increase the discharge head, although a small increase in capacity also results.

9.3 CENTRIFUGAL PUMPS[39]

The *centrifugal pump* (and its modifications) is the most widely used type of pumping equipment in water/wastewater operations. This type of pump is capable of moving high volumes of water/wastewater (and other liquids) in a relatively efficient manner. The centrifugal pump is very dependable, has relatively low maintenance requirements, and can be constructed out of a wide variety of construction materials. It is considered one of the most dependable systems available for water transfer.

9.3.1 CENTRIFUGAL PUMPS: DESCRIPTION

The centrifugal pump consists of a rotating element **(impeller)** sealed in a casing **(volute).** The rotating element is connected to a drive unit **(motor/engine)** that supplies the energy to spin the rotating element. As the impeller spins inside the volute casing, an area of low pressure is created in the center of the impeller. This low pressure allows the atmospheric pressure on the liquid in the supply tank to force the liquid up to the impeller. Because the pump will not operate if there is no low pressure zone created at the center of the impeller, it is important that the casing be sealed to prevent air from entering the casing. To ensure the casing is airtight, the pump employs some type of seal **(mechanical or conventional packing)** assembly at the point where the shaft enters the casing. This seal also includes lubrication, provided by either water, grease, or oil, to prevent excessive wear.

From a hydraulic standpoint, note the energy changes that occur in the moving water. As water enters the casing, the spinning action of the impeller imparts (transfers) energy to the water. This energy is transferred to the water in the form of increased speed or velocity. The liquid is thrown outward by the impeller into the volute casing where the design of the

[39] Much of the information contained in this section is adapted from training materials used by the U.S. Navy in its Class A Engineman Training Program, San Diego, California, and from Basic Maintenance Training Course, Onondaga County Department of Drainage and Sanitation, North Syracuse, New York, 1986.

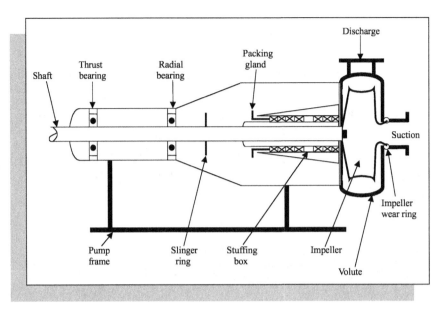

Figure 9.14
Centrifugal pump—major components.

casing (see Section 9.3.2) allows the velocity of the liquid to be reduced, which, in turn, converts the velocity energy **(velocity head)** to pressure energy **(pressure head).** The process by which this change occurs is described later. The liquid then travels out of the pump through the pump discharge. The major components of the centrifugal pump are shown in Figure 9.14.

> **Key Point:** A centrifugal pump is a pumping mechanism whose rapidly spinning impeller imparts a high velocity to the water that
> IMPORTANT enters then converts that velocity to pressure upon exit.

9.3.1.1 CENTRIFUGAL PUMP TERMINOLOGY

In order to understand centrifugal pumps and their operation, one must understand the terminology associated with centrifugal pumps.

 Base plate—the foundation under a pump. It usually extends far enough to support the drive unit. The base plate is often referred to as the pump frame.

💧 *Bearings*—devices used to reduce friction and to allow the shaft to rotate easily. Bearings may be sleeve, roller, or ball.

- **Thrust bearing**—in a single suction pump, it is the bearing located nearest the motor, farthest from the impeller. It takes up the major thrust of the shaft, which is opposite from the discharge direction.

- **Radial (line) bearing**—In a single suction pump, it is the one closest to the pump. It rides free in its own section and takes up and down stresses.

Note: *In most cases, where pump and motor are constructed on a common shaft (no coupling), the bearings will be part of the motor assembly.*

IMPORTANT

💧 *Casing*—the housing surrounding the rotating element of the pump. In the majority of centrifugal pumps, this casing can also be called the **volute.**

- **Split casing**—a pump casing that is manufactured in two pieces fastened together by means of bolts. Split-casing pumps may be vertically (perpendicular to the shaft direction) split or horizontally (parallel to the shaft direction) split (see Figure 9.15).

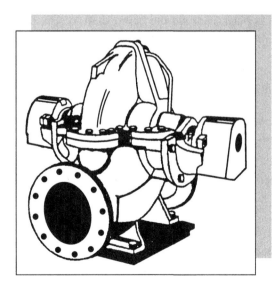

Figure 9.15
Split case centrifugal pump.

- **Coupling**—device to join the pump shaft to the motor shaft. If pump and motor are constructed on a common shaft, it is called a **close coupled** arrangement.

- **Extended shaft**—pump is constructed on one shaft and must be connected to the motor by a coupling.

- **Frame**—the housing that supports the pump bearing assemblies. In an end suction pump, it may also be the support for the pump casing and the rotating element.

- **Impeller**—the rotating element in the pump that actually transfers the energy from the drive unit to the liquid. Depending on the pump application, the impeller may be open, semi-open, or closed. It may also be single or double suction.

- **Impeller eye**—the center of the impeller; the area that is subject to lower pressures due to the rapid movement of the liquid to the outer edge of the casing.

- **Prime**—filling the casing and impeller with liquid. If this area is not completely full of liquid, the centrifugal pump will not pump efficiently.

- **Seals**—devices used to stop the leakage of air into the inside of the casing around the shaft.

- **Packing**—material that is placed around the pump shaft to seal the shaft opening in the casing and prevent air leakage into the casing.

- **Stuffing box**—the assembly located around the shaft at the rear of the casing. It holds the packing and lantern ring.

- **Lantern ring**—also known as the **seal cage,** it is positioned between the rings of packing in the stuffing box to allow the introduction of a lubricant (water, oil, or grease) onto the surface of the shaft to reduce the friction between the packing and the rotating shaft.

- **Gland**—also known as the **packing gland,** it is a metal assembly that is designed to apply even pressure to the packing to compress it tightly around the shaft.

- **Mechanical seal**—a device consisting of a stationary element, a rotating element, and a spring to supply force to hold the two elements together. Mechanical seals may be either single or double units.

- **Shaft**—the rigid steel rod that transmits the energy from the motor to the pump impeller. Shafts may be either vertical or horizontal.

- **Shaft sleeve**—a piece of metal tubing placed over the shaft to protect the shaft as it passes through the packing or seal area. In some cases, the sleeve may also help to position the impeller on the shaft.

- **Shut-off head**—the head or pressure at which the centrifugal pump will stop discharging. It is also the pressure developed by the pump when it is operated against a closed discharge valve. This is also known as a **cut-off head.**

- **Shroud**—the metal plate that is used to either support the impeller vanes (open or semi-open impeller) or to enclose the vanes of the impeller (closed impeller).

- **Slinger ring**—a device to prevent pumped liquids from traveling along the shaft and entering the bearing assembly. A slinger ring is also called a **deflector.**

- **Wearing rings**—devices that are installed on stationary or moving parts within the pump casing to protect the casing and/or the impeller from wear due to the movement of liquid through points of small clearances.

- **Impeller ring**—a wearing ring installed directly on the impeller.

- **Casing ring**—a wearing ring installed in the casing of the pump. A casing ring is also known as the suction head ring.

- **Stuffing box cover ring**—a wearing ring installed at the impeller in an end suction pump to maintain the impeller clearances and to prevent casing wear.

9.3.2 CENTRIFUGAL PUMP THEORY

The volute-cased centrifugal pump (see Figure 9.16) provides the pumping action necessary to transfer liquids from one point to another. First, the drive unit (usually an electric motor) supplies energy to the pump impeller to make it spin. This energy is then transferred to the water by the impeller. The vanes of the impeller spin the liquid toward the outer edge of the impeller at a high rate of speed or velocity. This action is very similar to that which would occur when a bucket full of water with a small hole in the bottom is attached to a rope and spun. When sitting still, the water in the bucket will drain out slowly. However, when the bucket is spinning, the water will be forced through the hole at a much higher rate of speed.

Centrifugal pumps may be single stage, having a single impeller, or they may be multiple stage, having several impellers through which the fluid flows in series. Each impeller in the series increases the pressure of the fluid at the pump discharge. Pumps may have 30 or more stages in extreme cases. In centrifugal pumps, a correlation of pump capacity, head, and speed at optimum efficiency is used to classify the pump impellers with respect to their specific geometry. This correlation is called **specific speed** (see Section 9.2.12) and is an important parameter for analyzing pump performance.[40]

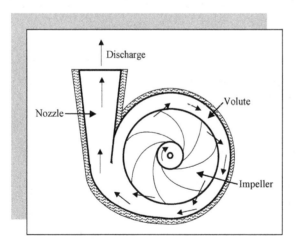

Figure 9.16
Cross-sectional diagram showing the features of a centrifugal pump.

[40] Garay, P. N., *Pump Application Desk Book.* Lilburn, GA: The Fairmont Press, Inc., p. 22, 1990.

The volute of the pump is designed to convert velocity energy to pressure energy. As a given volume of water moves from one cross-sectional area to another with the volute casing, the velocity or speed of the water changes proportionately. The volute casing has a cross-sectional area, which is extremely small at the point in the case that is farthest from the discharge (see Figure 9.16). This area increases continuously to the discharge. As this area increases, the velocity of the water passing through it decreases as it moves around the volute casing to the discharge point.

As the velocity of the water decreases, the velocity head decreases, and the energy is converted to pressure head. There is a direct relationship between the velocity of the water and the pressure it exerts. Therefore, as the velocity of the water decreases, the excess energy is converted to additional pressure (pressure head). This pressure head supplies the energy to move the water through the discharge piping.

9.3.3 PUMP CHARACTERISTICS

The centrifugal pump operates on the principle of an energy transfer and, therefore, has certain definite characteristics that make it unique. The type and size of the impeller limit the amount of energy that can be transferred to the water, the characteristics of the material being pumped, and the total head of the system through which the water is moving. For any one centrifugal pump, there is a definite relationship between these factors along with head (capacity), efficiency, and brake horsepower.

9.3.3.1 HEAD (CAPACITY)

As might be expected, the capacity of a centrifugal pump is directly related to the total head of the system. If the total head of the system is increased, the volume of the discharge will be reduced proportionately. As the head of the system increases, the capacity of the pump will decrease proportionately until the discharge stops. The head at which the discharge no longer occurs is known as the **cut-off head.**

As pointed out earlier, the total head includes a certain amount of energy to overcome the friction of the system. This friction head can be greatly affected by the size and configuration of the piping and the condition of the system's valving. If the control valves on the system are closed partially, the friction head can increase dramatically. When this happens, the total head increases, and the capacity or volume discharged by the pump decreases. In many cases, this method is employed to reduce the discharge of a centrifugal pump. It should be noted, however, that this does increase the load on the pump and drive system, causing additional energy requirements and additional wear.

The total closure of the discharge control valve increases the friction head to the point where all the energy supplied by the pump is consumed in the friction head and is not converted to pressure head. Consequently, the pump exceeds its cut-off head, and the pump discharge is reduced to zero. Again, it is important to note that, although the operation of a centrifugal pump against a closed discharge may not be hazardous (as with other types of pumps), it should be avoided because of the excessive load placed on the drive unit and pump. On occasion, the pump can produce pressure higher than the pump discharge piping can withstand. Whenever this occurs, the discharge piping may be severely damaged by the operation of the pump against a closed or plugged discharge.

9.3.3.2 EFFICIENCY

Every centrifugal pump will operate with varying degrees of efficiency over its entire capacity and head ranges. The important factor in selecting a centrifugal pump is to select a unit that will perform near its maximum efficiency in the expected application.

9.3.3.3 BRAKE HORSEPOWER REQUIREMENTS

In addition to the head capacity and efficiency factors, most pump literature includes a graph showing the amount of energy in horsepower that must be supplied to the pump to obtain optimal performance.

9.3.4 THE CENTRIFUGAL PUMP: ADVANTAGES AND DISADVANTAGES

Centrifugal pumps have become one of the most widely used types of pumps because of the several advantages it offers:

◆ **Construction**—The pump consists of a single rotating element and simple casing, which can be constructed using a wide assortment of materials. If the fluids to be pumped are highly corrosive, the pump parts that are exposed to the fluid can be constructed of lead or other material that is not likely to corrode. If the fluid being pumped is highly abrasive, the internal parts can be made of abrasion-resistant material or coated with a protective material.

   Also, the simple design of a centrifugal pump allows the pump to be constructed in a variety of sizes and configurations. No other pump currently available has the range of capacities or applications available through the use of the centrifugal pump.

◆ **Operation**—Simple and quiet best describes the operation of a centrifugal pump. An operator-in-training with a minimum amount of experience may be capable of operating facilities that use centrifugal-type pumps. Even when improperly operated, the centrifugal pump's rugged construction allows it to operate (in most cases) without major damage.

◆ **Maintenance**—The amount of wear on a centrifugal pump's moving parts is reduced and its operating life is extended because its moving parts are not required to be constructed to very close tolerances.

◆ **Pressure is self-limited**—Because of the nature of its pumping action, the centrifugal pump will not exceed a predetermined maximum pressure. Thus, if the discharge valve is suddenly closed, the pump cannot generate additional pressure that might result in damage to the system or could potentially result in a hazardous working condition. The power supplied to the impeller will only generate a specified amount of head (pressure). If a major portion of this head or pressure is consumed in overcoming friction or is lost as heat energy, the pump will have a decreased capacity.

- **Adaptable to high-speed drive systems**—Allows the use of high speed, high efficiency motors. In situations where the pump is selected to match a specific operating condition, which remains relatively constant, the pump drive unit can be used without the need for expensive speed reducers.

- **Small space requirements**—For most pumping capacities, the amount of space required for installation of the centrifugal-type pump is much less than that of any other type of pump.

- **Fewer moving parts**—Rotary rather than reciprocating motion employed in centrifugal pumps reduces space and maintenance requirements due to the fewer number of moving parts required.

Although the centrifugal pump is one of the most widely used pumps, it does have a few disadvantages:

- **Additional equipment needed for priming**—The centrifugal pump can be installed in a manner that will make it self-priming, but it is not capable of drawing water to the pump impeller unless the pump casing and impeller are filled with water. This can cause problems, because if the water in the casing drains out, the pump would cease pumping until it is refilled.

  Therefore, it is normally necessary to start a centrifugal pump with the discharge valve closed. The valve is then gradually opened to its proper operating level. Starting the pump against a closed discharge valve is not hazardous provided the valve is not left closed for extended periods.

- **Air leaks affect pump performance**—Air leaks on the suction side of the pump can cause reduced pumping capacity in several ways. If the leak is not serious enough to result in a total loss of prime, the pump may operate at a reduced head or capacity due to air mixing with the water. This causes the water to be lighter than normal and reduces the efficiency of the energy transfer process.

- **Narrow range of efficiency**—Centrifugal pump efficiency is directly related to the head capacity of the pump. The highest performance efficiency is available for only a very small section of the head-capacity range. When the pump is operated outside of this optimum range, the efficiency may be greatly reduced.

● **Pump may run backwards**—If a centrifugal pump is stopped without closing the discharge line, it may run backwards, because the pump does not have any built-in mechanism to prevent flow from moving through the pump in the opposite direction (i.e., from discharge side to suction). If the discharge valve is not closed or the system does not contain the proper check valves, the flow that was pumped from the supply tank to the discharge point will immediately flow back to the supply tank when the pump shuts off. This results in increased power consumption due to the frequent startup of the pump to transfer the same liquid from supply to discharge.

Note: *It is sometimes difficult to tell whether a centrifugal pump is running forward or backwards because it appears and sounds like it is operating normally when operating in reverse.*

IMPORTANT

● **Pump speed is difficult to adjust**—centrifugal pump speed cannot usually be adjusted without the use of additional equipment, such as speed-reducing or speed-increasing gears or special drive units.

Because the speed of the pump is directly related to the discharge capacity of the pump, the primary method available to adjust the output of the pump other than a valve on the discharge line is to adjust the speed of the impeller. Unlike some other types of pumps, the delivery of the centrifugal pump cannot be adjusted by changing some operating parameter of the pump.

9.3.5 CENTRIFUGAL PUMP APPLICATION

The centrifugal pump is probably the most widely used pump available at this time because of its simplicity of design and wide ranging diversity (it can be adjusted to suit a multitude of applications). Proper selection of the pump components (impeller, casing, etc.) and construction materials can produce a centrifugal pump capable of transporting not only water but also other materials ranging from material/chemical slurries to air (centrifugal blowers). To attempt to list all of the various applications for the centrifugal pump would exceed the limitations of this guidebook. Therefore, the discussion of pump applications is limited to those that frequently occur in water/wastewater operations.

- *Large volume pumping*—In water/wastewater operations, the primary use of centrifugal pumps is large volume pumping. In large volume pumping, generally low speed, moderate head, vertically shafted pumps are used. Centrifugal pumps are well suited for water/wastewater system operations because they can be used in conditions where high volumes are required and a change in flow is not a problem. As the discharge pressure on a centrifugal pump is increased, the quantity of water/wastewater pumped is reduced. Also, centrifugal pumps can be operated for short periods with the discharge valve closed.

- *Non-clog pumping*—Specifically designed centrifugal pumps using closed impellers with, at most, two to three vanes. It is usually designed to pass solids or trash up to 3" in diameter.

- *Dry pit pump*—Depending on the application, may be either a large volume pump or a non-clog pump. It is located in a dry pit that shares a common wall with the wet well. This pump is normally placed in such a position to ensure that the liquid level in the wet well is sufficient to maintain the pump's prime.

- *Wet pit or submersible pump*—Usually a non-clog type pump that can be submerged, with its motor, directly in the wet well. In a few instances, the pump may be submerged in the wet well while the motor remains above the water level. In these cases, the pump is connected to the motor by an extended shaft.

- *Underground pump stations*—Using the wet well-dry well design, the pumps are located in an underground facility. Wastes are collected in a separate wet well, then pumped upward and discharged into another collector line or manhole. This system normally uses a non-clog type pump and is designed to add sufficient head to water/waste flow to allow gravity to move the flow to the plant or the next pump station.

● *Recycle or recirculation pumps*—Because the liquids being transferred by the recycle or recirculation pump normally do not contain any large solids, the use of the non-clog type centrifugal pump is not always required. A standard centrifugal pump may be used to recycle trickling filter effluent, return activated sludge, or digester supernatant.

● *Service water pumps*—The wastewater plant effluent may be used for many purposes, such as to clean tanks, water lawns, provide the water to operate the chlorination system, and backwash filters. Because the plant effluent used for these purposes is normally clean, the centrifugal pumps used closely parallel the units used for potable water. In many cases, the double suction, closed impeller, or turbine type pump will be used.

9.4 PUMP CONTROL SYSTEMS

"Pump operations usually control only one variable: flow, pressure or level. All pump control systems have a measuring device that compares a measured value with a desired one. This information relays to a control element that makes the changes. . . . The user may obtain control with manually operated valves or sophisticated microprocessors. Economics dictate the accuracy and complication of a control system."[41]

Most centrifugal pumps require some form of pump control system. The only exception to this practice is when the plant pumping facilities are designed to operate continuously at a constant rate of discharge. The typical pump control system includes a sensor to determine when the pump should be turned on or off and the electrical/electronic controls to actually start and stop the pump.

The control systems currently available for the centrifugal pump range from a very simple on-off float control to an extremely complex system capable of controlling several pumps in sequence.

The following sections briefly describe the operation of various types of control devices/systems used with centrifugal pumps.

---

[41] Wahren, U., *Practical Introduction to Pumping Technology.* Houston: Gulf Publishing Company, p. 128, 1997.

9.4.1 FLOAT CONTROL

Currently, the *float control system* is the simplest of the centrifugal pump controls (see Figure 9.17). In the float control system, the float rides on the water's surface in the well, storage tank, or clear well, attached to the pump controls by a rod with two collars. One collar activates the pump when the liquid level in the well or tank reaches a preset level, and a second collar shuts the pump off when the level in the well reaches a minimum level. This type of control system is simple to operate and relatively inexpensive to install and maintain. The system has several disadvantages.

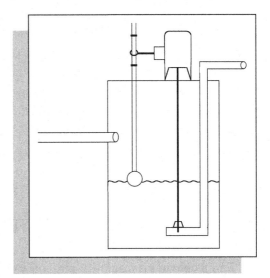

Figure 9.17
Float system for pump motor control.

The system operates at one discharge rate. This can result in (1) extreme variations in the hydraulic loading on succeeding units and (2) long periods of non-operation due to low flow periods or maintenance activities.

9.4.2 PNEUMATIC CONTROLS

Pneumatic control systems (also called *bubbler tube control systems*) are relatively simple systems that can be used to control one or more pumps. The system consists of an air compressor, a tube extending into the well, clear well or storage tank/basin, and pressure-sensitive switches with varying on/off set points and a pressure relief valve (see Figure 9.18).

The system works on the basic principle that measures the depth of the water in the well or tank by determining the air pressure that is necessary to just release a bubble from the bottom of the tube (see

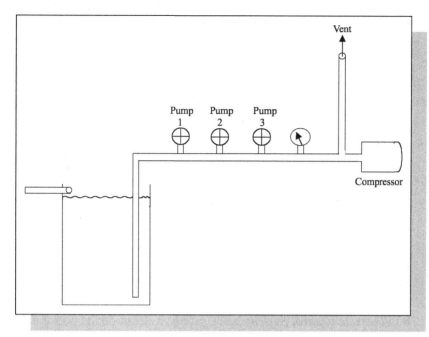

Figure 9.18
Pneumatic system for pump motor control.

Figure 9.18), hence, the name "bubbler tube." The air pressure required to force a bubble out of the tube is determined by the liquid pressure, which is directly related to the depth of the liquid (1 psi = 2.31 ft). By installing a pressure switch on the airline to activate the pump starter at a given pressure, the level of the water can be controlled by activating one or more pumps.

Installation of additional pressure switches with slightly different pressure settings allows several pumps to be activated in sequence. For example, the first pressure switch can be adjusted to activate a pump when the level in the well/tank is 3.8 ft (1.6 psi) and shut off at 1.7 ft (0.74 psi). If the flow into the pump well/tank varies greatly and additional pumps are available to ensure that the level in the well/tank does not exceed the design capacity, additional pressure switches may be installed. These additional pressure switches are set to activate a second pump when the level in the well/tank reaches a preset level (i.e., 4.5 ft/1.95 psi) and to cut off when the well/tank level is reduced to a preset level (i.e., 2.7 ft/1.2 psi). If the first pump's capacity is less than the rate of flow into the well/tank, the level of the well/tank continues to rise. Upon reaching the preset level

(4-ft level), it will activate the second pump. If necessary, a third pump can be added to the system set to activate at a third preset well/tank depth (4.6 ft/1.99 psi) and cut off at a preset depth (3.0 ft/1.3 psi).

The pneumatic control system is relatively simple with minimal operation and maintenance requirements. The major operational problem involved with this control system is the clogging of the bubbler tube. If, for some reason, the tube becomes clogged, the pressure on the system can increase and may activate all pumps to run even when the well/tank is low. This can result in excessive power consumption, which, in turn, may damage the pumps.

9.4.3 ELECTRODE CONTROL SYSTEMS

The *Electrode Control System* uses a probe or electrode to control the pump on and off cycle. A relatively simple control system, it consists of two electrodes extended into the clear well, storage tank, or basin. One electrode is designed to activate the pump starter when it is submerged in the water; the second electrode extends deeper into the well/tank and is designed to open the pump circuit when the water drops below the electrode (see Figure 9.19).

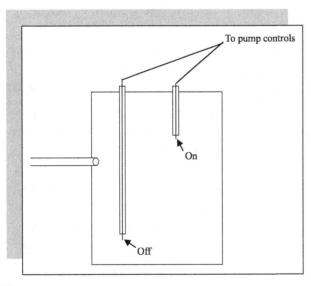

Figure 9.19
Electrode system for pump motor control.

The major maintenance requirement of this system is keeping the electrodes clean.

 Important Point: *Because the Electrode Control System uses two separate electrodes, the unit may be locked into an on-cycle or off-cycle depending on which electrode is involved.*

9.4.4 OTHER CONTROL SYSTEMS

Several other systems that use electrical energy are available for control of the centrifugal pump. These include a *tube-like device* that has several electrical contacts mounted inside (see Figure 9.20). As the water level rises in the clear well, storage tank, or basin, the water rises in the tube making contact with the electrical contacts and activating the motor starter. Again, this system can be used to activate several pumps in series by installing several sets of contact points.

As the water level drops in the well/tank, the level in the tube drops below a second contact that deactivates the motor and stops the pumping.

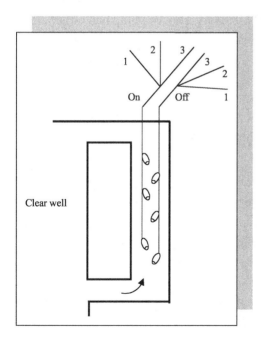

Figure 9.20
Electrical contacts for pump motor control.

Another control system uses a *mercury switch* (or a similar-type of switch) enclosed in a protective capsule. Again, two units are required per pump. One switch activates the pump when the liquid level rises, and the second switch shuts the pump off when the level reaches the desired minimum depth.

9.4.4.1 ELECTRONIC SYSTEMS

Several centrifugal pump control systems are available that use electronic systems for control of pump operation. A brief description of some of these systems is provided in the sections that follow.

9.4.4.1.1 FLOW EQUALIZATION SYSTEM

In any multiple-pump operation, the flow delivered by each pump will vary due to the basic hydraulic design of the system. To obtain equal loads on each pump when two or more are in operation, the flow equalization system electronically monitors the delivery of each pump and adjusts the speed of the pumps to obtain similar discharge rates for each pump.

9.4.4.1.2 SONAR OR OTHER TRANSMISSION TYPE CONTROLLERS

A *sonar* or *low-level radiation system* can be used to control centrifugal pumps. This type of system uses a transmitter and a receiver to locate the level of the water in a tank, clear well, or basin. When the level reaches a predetermined set point, the pump is activated, and when the level is reduced to a predetermined set point, the pump is shut off. Basically, the system is very similar to a radar unit. The transmitter sends out a beam that travels to the liquid, bounces off the surface, and returns to the receiver. The time required for this is directly proportional to the distance from the liquid to the instrument. The electronic components of the system can be adjusted to activate the pump when the time interval corresponds to a specific depth in the well or tank. The electronic system can also be set to shut off the pump when the time interval corresponds to a preset minimum depth.

9.4.4.1.3 MOTOR CONTROLLERS

There are several types of controllers available not only to protect the motor from overloads but also from short-circuit conditions. Many motor controllers also function to adjust motor speed to increase or decrease the discharge rate for a centrifugal pump. This type of control may use one of the previously described controls to start and stop the pump and, in some cases, adjust the speed of the unit. As the depth of the water in a well or tank increases, the sensor automatically increases the speed of the motor in predetermined steps to the maximum design speed. If the level continues to increase, the sensor may be designed to activate an additional pump.

9.5 PROTECTIVE INSTRUMENTATION

Protective instrumentation of some type is normally employed in pump or motor installation. (Note: The information provided in this section applies to the centrifugal pump as well as to many other types of pumps.)

Protective instrumentation for centrifugal pumps (or most other types of pumps) is dependent on pump size, application, and the amount of operator supervision. That is, pumps less than 500 hp often only come with pressure gauges and temperature indicators. These gauges or transducers may be mounted locally (on the pump itself) or remotely (in suction and discharge lines immediately upstream and downstream of the suction and discharge nozzles). If transducers are employed, readings are typically displayed and taken (or automatically recorded) at a remote operating panel or control center.

9.5.1 TEMPERATURE DETECTORS[42]

Resistance temperature devices (RTDs) and *thermocouples* (see Figure 9.21) are commonly used as temperature detectors on the pump prime movers (motors) to indicate temperature problems. In some cases, dial thermometers, armored glass-stem thermometers, or bimetallic-actuated temperature indicators are used. The device employed typically monitors temperature variances that may indicate a possible source of trouble. On electric motors greater than 250 hp, RTD elements are used to monitor temperatures in stator winding coils. Two RTDs per phase is standard. One RTD element is usually installed in the shoe of the loaded area (see Figure 9.21) employed on journal bearings in pumps and motors. Normally, tilted-pad thrust bearings have an RTD element in the active, as well as the inactive, side.

RTDs are used when remote indication, recording, or automatic logging of temperature readings is required. Because of their smaller size, RTDs provide more flexibility in locating the measuring device near the measuring point.

When dial thermometers are installed, they monitor oil thrown from bearings. Sometimes, temperature detectors also monitor bearings with water-cooled jackets to warn against water supply failure. Pumps with heavy wall casing may also have casing temperature monitors.

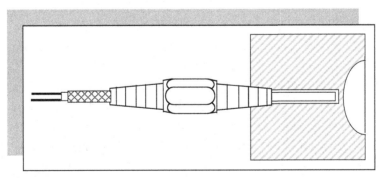

Figure 9.21
Thermocouple installation in journal bearing.

[42] From Grimes, A. S., Supervisory and Monitoring Instrumentation. In *Pump Handbook*, Karassik, I. J. et al. (eds.). New York: McGraw-Hill Book Company, pp. 8.4–8.8, 1976.

9.5.2 VIBRATION MONITORS

Vibration sensors are available to measure either bearing vibration or shaft vibration direction directly. Direct measurement of shaft vibration is desirable for machines with stiff bearing supports where bearing-cap measurements will be only a fraction of the shaft vibration. Wahren[43] points out that pumps and motors 1000 hp and larger may have the following vibration monitoring equipment:

● a seismic pickup with double set points installed on the pump outboard housing

● proximators with X-Y vibration probes complete with interconnecting co-axial cables at each radial and thrust journal bearing

● key-phasor with proximator and interconnecting co-axial cables

9.5.3 SUPERVISORY INSTRUMENTATION

Supervisory instruments are used to monitor the routine operation of pumps, their prime movers, and their accessories in order to sustain a desired level of reliability and performance. Generally, these instruments are not used for accurate performance tests or for automatic control, although they may share connections or functions.

Supervisory instruments consist of annunciators and alarms that provide operators with warnings of abnormal conditions that, unless corrected, will cause pump failure. Annunciators used for both alarm and pre-alarm have both visible and audible signals.

9.6 CENTRIFUGAL PUMP MODIFICATIONS

The centrifugal pump can be modified to meet the needs of several different applications. If there is a need to produce higher discharge heads,

---

[43] Wahren, U. *Practical Introduction to Pumping Technology.* Houston, TX: Gulf Publishing Company, p. 137, 1997.

the pump may be modified to include several additional impellers. If the material being pumped contains a large amount of material that could clog the pump, the pump construction may be modified to remove a major portion of the impeller from direct contact with material being pumped.

Although there are numerous modifications of the centrifugal pump available, the scope of this text covers only those that have found wide application in the water distribution and wastewater collection and treatment fields. Modifications to be presented in this section include

- submersible pumps

- recessed impeller or vortex pumps

- turbine pumps

9.6.1 SUBMERSIBLE PUMPS

The *submersible pump* is, as the name suggests, placed directly in the wet well or groundwater well. It uses a waterproof electric motor located below the static level of the wet well/well to drive a series of impellers. In some cases, only the pump is submerged, while, in other cases, the entire pump-motor assembly is submerged. Figure 9.22 illustrates this system.

9.6.1.1 DESCRIPTION

The submersible pump may be either a close-coupled centrifugal pump or an extended shaft centrifugal pump. If the system is a close-coupled system (as shown in Figure 9.22), then both motor and pump are submerged in the liquid being pumped. Seals prevent water/wastewater from entering the inside of the motor, protecting the electric motor in a close-coupled pump from shorts and motor burnout.

In the extended shaft system, the pump is submerged while the motor is mounted above the pump wet well. In this situation, an extended shaft assembly must connect the pump and motor.

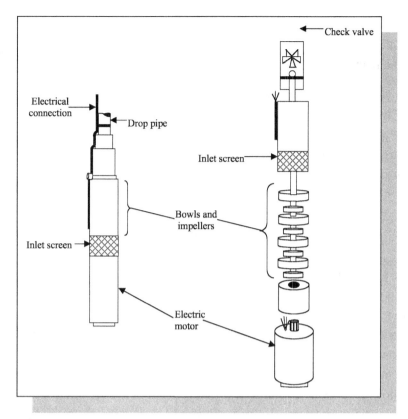

Figure 9.22
Submersible pump.

9.6.1.2 APPLICATIONS

The submersible pump has wide applications in the water/wastewater treatment industry. It generally can be substituted in any application of other types of centrifugal pumps. However, it has found its widest application in the distribution or collector system pump stations.

9.6.1.3 ADVANTAGES

In addition to the advantages discussed earlier for a conventional centrifugal pump, the submersible pump has additional advantages:

- It is located below the surface of the liquid, and there is less chance that the pump will lose its prime, develop air leaks on the suction side of the pump, or require initial priming.

- The pump or the entire assembly is located in the well/wet well, and there is less cost associated with the construction and operation of this system. There is no need to construct a dry well or a large structure to hold the pumping equipment and necessary controls.

9.6.1.4 DISADVANTAGES

The major disadvantage associated with the submersible pump is the lack of access to the pump or pump and motor. The performance of any maintenance requires either drainage of the wet well or extensive lift equipment to remove the equipment from the wet well or both. This may be a major factor in determining if a pump receives the attention it requires. Also, in most cases, all major maintenance on close-coupled submersible pumps must be performed by outside contractors due to the need to re-seal the motor to prevent leakage.

9.6.2 RECESSED IMPELLER OR VORTEX PUMPS

The *recessed impeller* or *vortex pump* uses an impeller that is either partially or wholly recessed into the rear of the casing (see Figure 9.23). The spinning action of the impeller creates a vortex or whirlpool. This whirlpool increases the velocity of the material being pumped. As in other centrifugal pumps, this increased velocity is then converted to increased pressure or head.

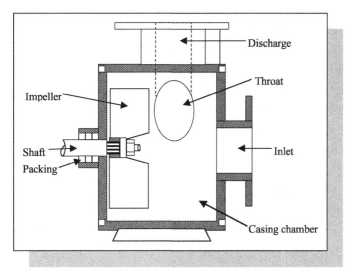

Figure 9.23
Schematic of a recessed impeller or vortex pump.

9.6.2.1 APPLICATIONS

The recessed impeller or vortex pump is used widely in applications where the liquid being pumped contains large amounts of solids or debris and slurries that could clog or damage the pump's impeller. It has found increasing use as a sludge pump in facilities that withdraw sludge continuously from their primary clarifiers.

9.6.2.2 ADVANTAGES

The major advantage of this modification is the increased ability to handle materials that would normally clog or damage the pump impeller. Because the majority of the flow does not come in direct contact with the impeller, there is much less potential for problems.

9.6.2.3 DISADVANTAGES

Because there is less direct contact between the liquid and the impeller, the energy transfer is less efficient. This results in somewhat higher power costs and limits the pump's application to low to moderate capacities.

Objects that might have clogged a conventional type centrifugal pump are able to pass through the pump. Although this is very beneficial in reducing pump maintenance requirements, it has, in some situations, allowed material to be passed into a less accessible location before becoming an obstruction. To be effective, the piping and valving must be designed to pass objects of a size equal to that which the pump will discharge.

9.6.3 TURBINE PUMPS

The turbine pump consists of a motor, drive shaft, a discharge pipe of varying lengths, and one or more impeller-bowl assemblies (see Figure 9.24). It is normally a vertical assembly where water enters at the bottom, passes axial through the impeller-bowl assembly where the energy transfer occurs, and then moves upward through additional impeller-bowl assemblies to the discharge pipe. The length of this discharge pipe will vary with the distance from the wet well to the desired point of discharge.

9.6.3.1 APPLICATION

Due to the construction of the turbine pump, the major applications have traditionally been for pumping of relatively clean water. The line shaft turbine pump has been used extensively for drinking water pumping, especially in those situations where water is withdrawn from deep wells. The main wastewater plant application has been pumping plant effluent back into the plant for use as service water.

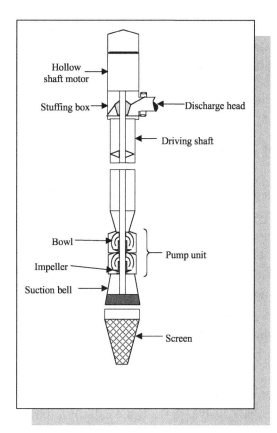

Figure 9.24
Vertical turbine pump.

9.6.3.2 ADVANTAGES

The turbine pump has a major advantage in the amount of head it is capable of producing. By installing additional impeller-bowl assemblies, the pump is capable of even greater production. Moreover, the turbine pump has simple construction and a low noise level and is adaptable to several drive types—motor, engine, or turbine.

9.6.3.3 DISADVANTAGES

High initial cost and high repair costs are two of the major disadvantages of turbine pumps. In addition, the presence of large amounts of solids within the liquid being pumped can seriously increase the amount of maintenance the pump requires. Consequently, the unit has not found widespread use in any other situation other than service water pumping.

9.7 POSITIVE-DISPLACEMENT PUMPS

Positive-displacement pumps force or displace water through the pumping mechanism. Most have a reciprocating element that draws water into the pump chamber on one stroke and pushes it out on the other. Unlike centrifugal pumps that are meant for low-pressure, high-flow applications, positive-displacement pumps can achieve greater pressures, but are slower moving, low-flow pumps.

Other positive-displacement pumps include the piston pump, diaphragm pump, and peristaltic pump, which are the focus of this discussion. In the water/wastewater industry, positive-displacement pumps are most often found as chemical feed pumps.

It is important to remember that positive-displacement pumps **cannot** be operated against a closed discharge valve. As the name indicates, something must be displaced with each stroke of the pump. Closing the discharge valve can cause rupturing of the discharge pipe, the pump head, the valve, or some other component.

9.7.1 PISTON PUMP OR RECIPROCATING PUMP

The *piston* or *reciprocating pump* is one type of positive-displacement pump. This pump works just like the piston in an automobile engine—on the intake stroke, the intake valve opens, filling the cylinder with liquid. As the piston reverses direction, the intake valve is pushed closed, and the discharge valve is pushed open; the liquid is pushed into the discharge pipe. With the next reversal of the piston, the discharge valve is pulled closed, the intake valve is pulled open, and the cycle repeats.

A piston pump is usually equipped with an electric motor and a gear and cam system that drives a plunger connected to the piston. Just like an automobile engine piston, the piston must have packing rings to prevent leakage and must be lubricated to reduce friction. Because the piston is in contact with the liquid being pumped, only good grade lubricants can be used when pumping materials that will be added to drinking water. The valves must be replaced periodically as well.

9.7.2 DIAPHRAGM PUMP

A *diaphragm pump* is composed of a chamber used to pump the fluid; a diaphragm that is operated by either electric or mechanical means; and two valve assemblies—a suction and a discharge valve assembly (see Figure 9.25).

A diaphragm pump is a variation of the piston pump in which the plunger is isolated from the liquid being pumped by a rubber or synthetic diaphragm. As the diaphragm is moved back and forth by the plunger, liquid is pulled into and pushed out of the pump. This arrangement provides better protection against leakage of the liquid being pumped and allows the use of lubricants that otherwise would not be permitted. Care must be taken to ensure that diaphragms are replaced before they rupture. Diaphragm pumps are appropriate for discharge pressures up to about 125 psi, but do not work well if they must lift liquids more than about four feet.

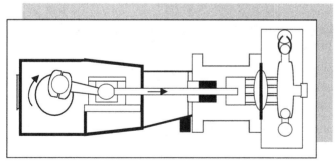

Figure 9.25
Diaphragm pump.

Diaphragm pumps are frequently used for chemical feed pumps. By adjusting the frequency of the plunger motion and the length of the stroke, extremely accurate flow rates can be metered. The pump may be driven hydraulically by an electric motor or by an electronic driver in which the plunger is operated by a solenoid. Electronically driven metering pumps are extremely reliable (few moving parts) and inexpensive.

9.7.3 PERISTALTIC PUMPS

Peristaltic pumps (sometimes called tubing pumps) use a series of rollers to compress plastic tubing to move the liquid through the tubing. A rotary gear turns the rollers at a constant speed to meter the flow. Peristaltic pumps are mainly used as chemical feed pumps.

The flow rate is adjusted by changing the speed the roller-gear rotates (to push the waves faster) or by changing the size of the tubing (so there is more liquid in each wave). As long as the right type of tubing is used, peristaltic pumps can operate at discharge pressures up to 100 psi. Note that the tubing must be resistant to deterioration from the chemical being pumped. The principle item of maintenance is the periodic replacement of the tubing in the pump head. There are no check valves or diaphragms in this type of pump.

REFERENCES

Basic Maintenance Training Course, Onondaga County Department of Drainage and Sanitation. North Syracuse, NY: 1986.

Basic Science Concepts and Applications: Principles and Practices of Water Supply Operations, 2nd ed. Denver, CO: American Water Works Association, 1995.

Fleet Training Center, U.S. Navy. Class A Engineman Training Program, 1963.

Garay, P. N., *Pump Application Desk Book.* Lilburn, GA: The Fairmont Press, Inc., p. 1, 1990.

Grimes, A. S., Supervisory and Monitoring Instrumentation. In *Pump Handbook,* Karassik, I. J. et al. (eds.). New York: McGraw-Hill Book Company, 1976.

Hauser, B. A., *Hydraulics for Operators.* Boca Raton, FL: Lewis Publishers, 1993.

Hauser, B. A., *Practical Hydraulics Handbook,* 2nd ed. Boca Raton, FL: Lewis Publishers, 1996.

Hydraulic Institute Complete Pump Standards, 4th ed. Cleveland: Hydraulic Institute, 1994.

The Hydraulic Institute Engineering Data Book, 2nd ed. Cleveland: Hydraulic Institute, 1990.

Lindeburg, M. R., *Civil Engineering Reference Manual,* 4th ed. San Carlos, CA: Professional Publications, Inc., 1986.

Spellman, F. R., *The Science of Water.* Lancaster, PA: Technomic Publishing Co., Inc., 1997.

Spellman, F. R., *The Handbook for Waterworks Operator Certification, Volume 2: Intermediate Level.* Lancaster, PA: Technomic Publishing Co., Inc., 2000.

Texas Manual. *Manual of Water Utility Operations,* 8th ed. Texas Utilities Association, 1988.

Wahren, U., *Practical Introduction to Pumping Technology.* Houston: Gulf Publishing Company, 1997.

Water Treatment Operators Short Course. Blacksburg, VA: Virginia Polytechnic Institute and State University (Virginia Tech), 1999.

Self-Test

9.1 Define static discharge head.

9.2 If a pump is selected based on the _____ distance, the water must be _____, or the pump will not provide sufficient _____ to cause a discharge from the line.

9.3 _____ head is the amount of energy a pump supplies to keep the liquid moving at a given speed.

9.4 The flow through a 6-inch diameter pipe is 5 MGD. What is the velocity of the flow?

9.5 Flow is entering a pressure main. If the pressure in the main is 15 psi, what amount of head must be added to the total head of the pumping system to overcome the pressure in the main?

9.6 Name three major components of the centrifugal pump.

9.7 List three advantages of the centrifugal pump.

9.8 List three disadvantages of the centrifugal pump.

9.9 What is the purpose of the impeller?

9.10 What is the purpose of the stuffing box/packing assembly?

9.11 What is the purpose of the shaft sleeve?

9.12 Describe a simple float control system for a centrifugal pump and how it works.

9.13 Describe the control system that uses air pressure to activate the pump operation, and briefly describe how it works.

9.14 Describe the recessed impeller type centrifugal pump.

9.15 Describe the submersible type centrifugal pump.

9.16 Define cavitation.

9.17 Define brake horsepower.

9.18 Explain the difference between hydraulic horsepower and brake horsepower.

9.19 What is NPSH?

9.20 Explain the difference between running pumps in series and in parallel.

Final Review Examination

Note: *The answers for the comprehensive examination are contained in Appendix B.*

IMPORTANT

10.1 Many of the water quality problems in water systems are due to contamination of the water _____ it leaves the treatment facility.

10.2 Define hydraulics.

10.3 One cubic foot of water weighs _____ lb.

10.4 One cubic foot of water contains _____ gallons.

10.5 Define head.

10.6 If the top of a tube is tightly capped and all of the air is removed from the sealed tube above the water surface, forming a _____ vacuum, the pressure on the water surface inside of the tube will be _____ psi.

10.7 What is the pressure at a point 15 ft below the surface of a reservoir?

10.8 A perfect vacuum plus atmospheric pressure of 14.2 psi will lift water how far?

10.9 If the gauge pressure in a water main is 100 psi, how far will the water rise in a tube connected to the main?

10.10 A pipe 12 inches in diameter has water flowing through it at 10 ft/second. What is the discharge in cfs (area of pipe is 0.785 ft^2)?

10.11 A pipe 12 inches in diameter is connected to a 6-inch diameter pipe. The velocity of the water in the 12-inch pipe is 3 fps. What is the velocity in the 6-inch pipe?

10.12 The flow of water in pipes is caused by the _____ applied behind it.

10.13 The flow in a pipe is retarded by _____ of the water against the _____ of the pipe.

10.14 Resistance to flow offered by pipe friction depends on the _____ of the pipe, the _____ of the pipe wall, and the number and type of _____ along the pipe.

10.15 The more water you try to pump through a pipe, the more _____ it will take to overcome the friction.

10.16 Friction loss increases as flow rate _____, pipe length _____, pipe diameter _____, pipe is _____, pipe interior becomes _____, and as bends, fittings, and valves are _____

10.17 A number called the _____ indicates the roughness of a pipe.

10.18 The higher the C factor, the _____ the pipe.

10.19 It is standard practice to calculate the head loss from fittings by substituting the _____ from published tables.

10.20 The line that connects the piezometric surfaces along a pipeline is known as _____ .

10.21 The HGL always slopes _____ in the direction of _____ in a pipe.

10.22 The static head + the friction head in a given pipe system = _____ .

10.23 The low water level in a well is 186 ft below the ground surface, and the highest water level in a storage tank is 68 ft above the ground at the well. What is the static head?

10.24 The energy in feet that a pump supplies to the fluid is known as _____ .

10.25 The total head in feet on the suction or supply side of a pump, when the supply is located below the center of the pump, is known as _____ .

10.26 List three centrifugal pump modifications.

10.27 What are the advantages of the recessed impeller?

10.28 What are the advantages of the submersible pump?

10.29 What is the main use of the line-shaft turbine in a wastewater treatment plant?

10.30 If the flow through a 2-ft diameter pipe is 8 MGD, the velocity is _____ .

10.31 If the flow in a pipe has a velocity of 2 fps and a pressure of 5 psi and the velocity of the flow increases to 4 fps, the pressure will be _____ .

10.32 What is the friction head in a system that uses 150 ft of 6-inch diameter pipe, when the velocity is 3 fps? The system's valving is equivalent to an additional 75 ft of pipe. Reference material indicates a roughness factor of 0.025 for this particular pipe and flow rate.

10.33 What is the velocity head for a system that has a velocity of 4 fps?

10.34 A pump is discharging to a pipe that is full of liquid under a pressure of 15 psi. The pump and piping system has a total head of 97 ft. How much additional head must the pump supply to overcome the line pressure?

10.35 The metal assembly that is designed to apply even pressure to the packing to compress it tightly around a pump shaft is called _____ .

10.36 A device designed to prevent pumped liquids from traveling along the shaft and entering the bearing assembly is called _____ .

10.37 The capacity of a centrifugal pump is directly related to the total _____ of the system.

10.38 The important factor in selecting a centrifugal pump is to select a unit that will perform near its maximum _____ in the expected application.

10.39 The heart of a centrifugal pump is the _____ .

10.40 The water in a tank weighs 800 pounds. How many gallons does it hold?

10.41 If a liquid weighs 45-lb/cu ft, how much does a five-gallon can of the liquid weigh?

10.42 Chlorine gas is 2.5 times heavier than air. What is the weight of a cubic foot of chlorine?

10.43 A material whose specific gravity is 1.3 is dropped from a pier into a lake. How deep will it sink?

10.44 A flow of 1 MGD occurs in an 8-inch diameter pipeline. What is the water velocity?

10.45 How many gallons of water can be stored in a pipeline 4 ft in diameter and 6 miles long?

10.46 The pressure gauge at the bottom of a standpipe reads 110 psi. What is the depth of the water in the standpipe?

10.47 What is the static pressure (psi) 4 miles beneath the ocean surface?

10.48 A flow of 1500 gpm takes place in a 12-inch pipe. What is the velocity head?

10.49 What should be the minimum weir height for measuring a flow of 1200 gpm with a 90° V-notch weir, if the flow is now moving at 4 ft/sec in a 2-ft wide rectangular channel?

10.50 A piston pump chamber is 12 inches in diameter and 24 inches deep. How many gallons of sludge are moved with each stroke?

10.51 Gravitational force gives an object _____ .

10.52 A push or pull applied against an object to move it is called a(n) _____ .

10.53 The density of a liquid is expressed in terms of

_____ .

10.54 The specific gravity of a liquid is determined by comparing the weight of the fluid to the weight of an equal volume of _____ at the same temperature.

10.55 When the pipe handling the water is too small for the velocity of the fluid, the flow becomes _____ .

Appendix A

Answers to Chapter Self-Tests

CHAPTER 2

2.1   660 ft³ × 7.48 gal/ft³ = 4937 gallons

2.2   (1) 160 lb − 125 lb = 35 lb;
(2) specific gravity = 160/35 = 4.57

2.3   (1) (8.34 lb/gal) (0.91) = 7.59 lb/gal;
(2) (1450 gal) (7.59 lb/gal) = 11,005 lb

2.4   85 ft × 0.433 psi/ft = 36.8 psi

2.5   Height in feet $= \dfrac{16 \text{ psi}}{0.433 \text{ psi/ft}} = 37$ feet (rounded)

2.6   Static Head = Discharge Elevation − Supply Elevation
2666 ft − 2130 ft = 536 ft

2.7   liquid, gas

2.8   force

2.9   specific gravity

2.10   water

2.11   pressure

2.12 $\dfrac{910 \text{ lb}}{8.34 \text{ lb/gal}} = 109.1$ gal

2.13   8.34 lb/gal water $\times$ 1.10 SG liquid = 9.17 lb/gal liquid

40 gal/min $\times$ 9.17 lb/gal = 367 lb/min

367 lb/min $\times$ 1440 min/day = 528,480 lb/day

2.14   115 psi $\times$ 2.31 ft/psi = 265.7 ft

2.15 $\dfrac{1,600,000 \text{ gal}}{7.48}$ = 213,904 cu ft (rounded)

Volume = 7.85 $\times$ d^2 $\times$ depth

213,904 = 0.785 $\times$ 110^2 $\times$ depth

$\dfrac{213,904}{9498.5}$ = depth

depth = 22.5 ft

2.16   pressure = weight $\times$ height

6400 psf = 62.4 lb/cu ft $\times$ height

height = 103 ft (rounded)

2.17   35 psi $\times$ 144 sq in/sq ft = 5040 psf

force = pressure $\times$ area

lb = 5040 psf $\times$ 1.6 sq ft

lb = 8064

CHAPTER 3

3.1   always constant

3.2   pressure due to the depth of water

3.3   In a pipeline, when flow goes from a larger section of pipe to a smaller section, one of the things that happens is that the flow velocity is increased as the water moves from a large to a smaller section of pipe. Because Q is constant and $Q = A \times V$, when A gets smaller, V must get larger; the product of $A \times V$ must always equal Q. Conversely, if the area of flow increases, the velocity of flow must decrease. This principle is referred to as continuity of flow.

3.4   the line that connects the piezometric surfaces along a pipeline

3.5   1500 gpm $\times$ 1440 min/day = 2,160,000 gpd
$$= 2.16 \text{ MGD}$$

2.16 MGD $\times$ 1.55 cfs/MGD = 3.35 cfs

$Q = A \times V$

$3.35 = 0.785 \times 1^2 V$

$4.27 \text{ ft/sec} = V$

$$V_h = \frac{V^2}{2g} = \frac{(4.27)^2}{64.4}$$

$V_h = 0.28 \text{ ft}$

3.6   Pressure Head = 110 psi $\times$ 2.31 ft/psi = 254.1 ft

3.7 $$\frac{5.00 \text{ mL/sec}}{3785 \text{ mL/gal}} = 0.0013^2 \text{ gal/sec}$$

$$\frac{0.0013 \text{ gal/sec}}{7.48 \text{ gal/cu ft}} = 0.00018 \text{ cfs}$$

$Q = A \times V$

$0.00018 = .785 \times .33^2 \times V$

$V = 0.002 \text{ ft/sec}$

$$V_h = \frac{V^2}{2g} = \frac{0.002^2}{64.4} = 6.2 \times 10^{-8} \text{ ft}$$

3.8 $Q = A \times V$

$1.46 = 0.785 \times 0.5^2 V$

$7.44 \text{ ft/sec} = V$

$$V_h = \frac{V^2}{2g} = \frac{7.44^2}{64.4}$$

$V_h = 0.86 \text{ ft}$

CHAPTER 4

4.1   dynamic, displacement

4.2   pressure

4.3   static head

4.4   cone of depression

4.5   pressure energy due to the velocity of the water.

4.6 a pumping condition where the eye of the impeller of the pump is above the surface of the water from which the pump is pumping

4.7 water level when pump is on

4.8 90 psi $\times$ 2.31 ft/psi = 208 ft (rounded)

DDH = Lift + Losses

208 = 80 + x

x = 128 ft

4.9 $\dfrac{\text{gpm} \times hd}{3960} = \text{whp}$

$\dfrac{1000 \times hd}{3960} = 10$

hd = 40 ft (rounded)

$\dfrac{40 \text{ ft}}{2.31 \text{ ft/psi}} = 17.2 \text{ psi}$

4.10 Drawdown = Pumping Water Level − Static Water Level

Drawdown = 600 − 500

Drawdown = 100 ft

CHAPTER 5

5.1 pressure loss due to friction, ft

5.2 the slope of the specific energy line

5.3 Flow is steady if at any one point the velocity of the water does not vary and the pressure is constant.

5.4 the head loss per foot

5.5 roughness, length, diameter, velocity

CHAPTER 6

6.1 Two pipes are equivalent if the head loss generated by the water velocity is the same in both pipes.

6.2 two or more pipes of different age, material, size, or length, laid side by side, with the flow splitting among them

6.3 two or more pipes of different age, material, or size, laid end to end

CHAPTER 7

7.1 Critical flow is flow at the critical depth and velocity. Critical flow minimizes the specific energy and maximizes discharge. Subcritical flow is flow at greater than the critical depth (less than the critical velocity). Supercritical flow is flow at less than the critical depth (greater than the critical velocity).

7.2 wetted area divided by wetted perimeter; a measure of the efficiency with which the conduit can transmit flow

7.3 the ratio of area in flow to the width of the channel at the fluid surface

7.4 energy, maximizes

7.5 subcritical flow

CHAPTER 8

8.1 $$\frac{900 \text{ gpm}}{60 \text{ sec/min} \times 7.48 \text{ gal/cu ft}} = 2.01 \text{ cfs}$$

$Q = A \times V$

$2.01 = 2 \times d \times 3$

$d = 0.34 \text{ ft}$

$Q = 2.5 \times h^{2.5}$

$2.01 = 2.5 \ h^{2.5}$

$h = 0.92 \ h$

$0.34 \text{ ft} + 0.92 \text{ ft} = 1.26 \text{ ft}$

8.2 $Q = 3.33 \times L \times h^{1.5}$

$110 = 3.33 \times 10 \times h^{1.5}$

$h = 2.2$ ft head of water $+ 3$ ft height of weir
  $= 5.2$ depth

8.3 $\dfrac{600 \text{ gpm}}{60 \text{ sec/min} \times 7.48 \text{ gal/cu ft}} = 1.3 \text{ cfs}$

$Q = 2.5 \times h^{2.5}$

$1.3 = 2.5\, h^{2.5}$

$h = 0.77 \text{ ft}$

8.4 1.2 MGD $\times$ 1.55 cfs/MGD = 1.86 cfs

$Q = 3.33 \times L \times h^{1.5}$

$1.86 = 3.33 \times 40 \times h^{1.5}$

$h = 0.058 \text{ ft}$

0.058 ft $\times$ 12 in/ft = 0.7 inches

8.5 rate, amount

8.6 simple construction; relatively inexpensive; no moving parts; wide application; transmitting instruments are external; low maintenance

8.7 eccentric, segmental

8.8 Venturi

8.9 flow nozzle

8.10 Pitot tube

8.11 ultrasonic flowmeter

8.12 weir, flume

CHAPTER 9

9.1 the total head in feet from the pump datum to the discharge point.

9.2 vertical, raised, energy

9.3 velocity

9.4 Velocity (v), fps $= \dfrac{\text{Flow } (Q), \text{ cfs}}{\text{cross-sectional area } (A), \text{ ft}^2}$

V, fps = 5 MGD $\times$ 1.55 cfs/MGD

V, fps = 39.5

9.5 Head, ft = pressure, psi × 2.31 ft/psi
Head, ft = 15 psi × 2.31 ft/psi
Head, ft = 34.6

9.6 shaft, impeller, volute case

9.7 operation is simple and quiet; self-limitation of pressure; small space requirements

9.8 not self-priming; high efficiency only over a narrow range; pump can run backwards

9.9 to transfer mechanical energy of the motor to velocity head by centrifugal force

9.10 It seals the pump at the point where pump shaft passes through the volute case. This prevents air leakage into and water leaking out of pump.

9.11 to protect the shaft from wear caused by friction as it passes through the stuffing box/packing assembly

9.12 The float control system consists of a float connected to a rod that is hooked to the motor control unit. The float rides on the surface of the water in the clear well/tank. As the water level rises, the float moves the rod up, which causes a collar on the rod to switch the pump on. The pump is switched off in a similar manner.

9.13 The pneumatic system consists of an air compressor, a tube extending into the clear well, tank, or basin, a set of pressure switches, and a pressure relief valve. It uses air pressure to sense the water depth in the well/tank and turns the pumps on and off.

9.14 It uses an impeller that is either partially or wholly recessed into the rear of the casing. The spinning action of the impeller causes a vortex or whirlpool. The whirlpool increases the velocity of material being pumped.

9.15 utilizes a waterproof electric motor to drive a centrifugal type of pump in either a close-coupled or extended shaft configuration

9.16 constant formation and implosion of vapor cavities in a moving liquid, causing damage to pipes, pumps, and fittings

9.17 power applied to the pump

9.18 Hydraulic horsepower is the power necessary to lift water to a required height. Brake horsepower is the power applied to the pump.

9.19 net positive suction head

9.20 Series pumps are arranged one after the other, to increase pumping head. Parallel pumps are set side by side and designed with separate suction, for increasing the pumped flow.

Appendix B

Answers to Chapter 10—Final Review Examination

10.1 after

10.2 the study of fluids at rest and in motion

10.3 62.4

10.4 7.48

10.5 the height of a column of water in feet

10.6 perfect, zero

10.7 $p = w \times h$
$p = 62.4 \text{ lb/ft}^3 \times 15 \text{ ft}$
$p = 936 \text{ psf}$

10.8 34 ft

10.9 100 psi $\times$ 2.31 ft/psi = 231 ft

10.10 $Q = V \times A$
$= 10 \text{ ft/sec} \times 0.785 \text{ ft}^2$
$= 7.85 \text{ ft}^3/\text{sec or cfs}$

10.11 Using equation $A_1 V_1 = A_2 V_2$
$$V_2 = \frac{(0.785 \text{ ft}^2) \times (3 \text{ ft/sec})}{(0.196 \text{ ft}^2)}$$
$= 12 \text{ ft/sec or fps}$

10.12 pressure

10.13 friction, inside

10.14 size/diameter, roughness, fittings

10.15 pressure

10.16 increases, increases, decreases, constricted, rougher, added

10.17 C factor

10.18 smoother

10.19 equivalent length of pipe

10.20 hydraulic grade line

10.21 downward, flow

10.22 total dynamic head (TDH)

10.23 186 ft + 68 ft = 254 ft

10.24 pump head

10.25 suction lift

10.26 submersible pump, recessed impeller, turbine pumps

10.27 ability to handle liquid with large amounts of solids or debris that could clog or damage most pump impellers

10.28 The pump maintains its prime well and doesn't have air leak problems.

10.29 pumping relatively clean water such as plant effluent for service water

10.30
$$\text{Velocity, fps} = \frac{8 \text{ MGD} \times 1.55 \text{ cfs/MGD}}{0.785 \times 2 \text{ ft} \times 2 \text{ ft}}$$

$$\text{Velocity, fps} = \frac{12.4 \text{ cfs}}{3.14 \text{ ft}^2}$$

Velocity, fps = 3.95 fps

10.31 $P_1 \times V_1 = P_2 \times V_2$
5 psi $\times$ 2 fps = $P_2 \times$ 4 fps

$$P_2 = \frac{5 \text{ psi } \times 2 \text{ fps}}{4 \text{ fps}}$$

$$P_2 = 10 \text{ psi}/4$$

$$P_2 = 2.5 \text{ psi}$$

10.32 Friction Head, ft

$$= \text{Roughness Factor} \times \frac{\text{length}}{\text{diameter}} \times \frac{\text{velocity}^2}{2g}$$

Friction Head, ft

$$= 0.025 \times \frac{(150 \text{ ft} + 75 \text{ ft})}{0.5 \text{ ft}} \times \frac{(3 \text{ fps})^2}{2 \times 32 \text{ ft/sec}^2}$$

$$\text{Friction Head, ft} = 0.025 \times \frac{225 \text{ ft}}{0.5 \text{ ft}} \times \frac{9 \text{ ft}^2/\text{s}^2}{64 \text{ ft/sec}^2}$$

Friction Head, ft $= 0.025 \times 450 \times 0.1406$ ft

Friction Head, ft $= 1.58$ ft

10.33 $$\text{Velocity Head, ft} = \frac{(\text{Velocity})^2}{2 \times \text{Acceleration due to gravity}}$$

$$\text{Velocity Head, ft} = \frac{(4 \text{ fps})^2}{2 \times 32 \text{ ft/sec}^2}$$

$$\text{Velocity Head, ft} = \frac{16 \text{ ft}^2/\text{sec}^2}{64 \text{ ft/sec}^2}$$

Velocity Head, ft $= 0.25$ ft

10.34 Head, ft = pressure, psi $\times$ 2.31 ft/psi

$$= 15 \text{ psi} \times 2.31 \text{ ft/psi}$$

$$= 34.65 \text{ ft}$$

10.35 gland or packing gland

10.36 slinger

10.37 head

10.38 efficiency

10.39 impeller

10.40 $\dfrac{800 \text{ lb}}{8.34 \text{ lb/gal}} = 95.9$ gal or 96 gal

10.41 $45 \text{ lb/ft}^3 \times \dfrac{1 \text{ ft}^3}{7.48 \text{ gal}} \times 5 \text{ gal} = 30.1 \text{ lb}$

10.42 Multiply by density of air (0.075 lb/cu ft)
0.075 lb/cu ft $\times$ 2.5 = 0.1875 lb/cu ft

10.43 all the way to the bottom, because it is 1.3 times heavier than the water

10.44 $Q = A \times V$
$1.55 = 0.785 \times 0.67^2 \ V$
$4.4 = V$
$V = 4.4 \text{ ft/sec}$

10.45 Volume = $0.785 \times 4^2 \times 5280 \times 6$
$V = 397,900 \times 7.48$
$V = 2,976,298 \text{ gallons}$

10.46 110 psi $\times$ 2.31 ft/psi = 254.1 ft

10.47 5280 ft/mile $\times$ 4 = 21,120 ft
21,120 ft/2.3 ft/psi = 9143 psi

10.48 1500 gpm $\times$ 1440 min/day = 2,160,000 gpd
$= 2.16 \text{ MGD}$

2.16 MGD $\times$ 1.55 cfs/MGD = 3.35 cfs
$Q = A \times V$
$3.35 = 0.785 \times 1^2 \times V$
$V = 4.27 \text{ ft/sec}$
$V_h = V^2/2g = (4.27 \text{ ft/sec})^2/(64.4 \text{ ft/sec})^2$
$V_h = 0.28 \text{ ft}$

10.49 $\dfrac{1200 \text{ gpm}}{60 \text{ sec/min} \times 7.48 \text{ gal/cu ft}} = 2.67 \text{ cfs}$

$Q = A \times V$
$2.67 = 2 \times d \times 4$
$d = 0.33 \text{ ft}$
$Q = 2.5 \times h^{2.5}$
$2.67 = 2.5 \times h^{2.5}$
$1.03 \text{ ft} = h$
0.33 ft + original depth 1.03 head on weir = 1.36 ft tall

10.50 Volume = $0.785 \times d^2 \times h$

$= 0.785 \times 1^2 \times 2$

$= 1.57$ cu ft $\times 7.48$ gal/cu ft

$= 11.7$ gallons

10.51 weight

10.52 force

10.53 specific gravity

10.54 water

10.55 turbulent

Index

Printed in the United States
by Baker & Taylor Publisher Services